THE ENERGETICS OF
DEVELOPMENT

THE ENERGETICS OF DEVELOPMENT

A Study of Metabolism in the Frog Egg

LESTER G. BARTH
Professor of Zoology in Columbia University

LUCENA J. BARTH
Research Associate in Embryology in Columbia University

Columbia University Press, New York, 1954

COLUMBIA BICENTENNIAL EDITIONS AND STUDIES

The Energetics of Development
BY LESTER G. BARTH AND LUCENA J. BARTH

New Letters of Berlioz, 1830–1868
TEXT WITH TRANSLATION, EDITED BY JACQUES BARZUN

On the Determination of Molecular Weights by Sedimentation and Diffusion
BY CHARLES O. BECKMANN AND OTHERS

LUIGI PIRANDELLO: *Right You Are*
TRANSLATED AND EDITED BY ERIC BENTLEY

The Sculpture of the Hellenistic Age
BY MARGARETE BIEBER

The Algebraic Theory of Spinors
BY CLAUDE C. CHEVALLEY

HENRY CARTER ADAMS: *Relation of the State to Industrial Action* AND *Economics and Jurisprudence*
EDITED BY JOSEPH DORFMAN

ERNST CASSIRER: *The Question of Jean-Jacques Rousseau*
TRANSLATED AND EDITED BY PETER GAY

The Language of Taxonomy
BY JOHN R. GREGG

Ancilla to Classical Reading
BY MOSES HADAS

JAMES JOYCE: *Chamber Music*
EDITED BY WILLIAM Y. TINDALL

Apokrimata: Decisions of Septimius Severus on Legal Matters
EDITED BY WILLIAM L. WESTERMANN AND A. ARTHUR SCHILLER

COPYRIGHT 1954, COLUMBIA UNIVERSITY PRESS, NEW YORK

Published in Great Britain, Canada, India, and Pakistan
by Geoffrey Cumberlege, Oxford University Press
London, Toronto, Bombay, and Karachi

Library of Congress Catalog Card Number: 54-6712

MANUFACTURED IN THE UNITED STATES OF AMERICA

GENERAL EDITOR'S PREFACE

THE MODERN UNIVERSITY has become a great engine of public service. Its faculty of Science is expected to work for our health, comfort, and defense. Its faculty of Arts is supposed to delight us with plays and exhibits and to provide us with critical opinions, if not to lead in community singing. And its faculty of Political Science is called on to advise government and laity on the pressing problems of the hour. It is unquestionably right that the twentieth-century university should play this practical role.

But this conspicuous discharge of social duties has the effect of obscuring from the public—and sometimes from itself—the university's primary task, the fundamental work upon which all the other services depend. That primary task, that fundamental work, is Scholarship. In the laboratory this is called pure science; in the study and the classroom, it is research and teaching. For teaching no less than research demands original thought, and addressing students is equally a form of publication. Whatever the form or the medium, the university's power to serve the public presupposes the continuity of scholarship; and this in turn implies its encouragement. By its policy, a university may favor or hinder the birth of new truth. This is the whole meaning of the age-old struggle for academic freedom, not to mention the age-old myth of academic retreat from the noisy world.

Since these conditions of freedom constitute the main theme of Columbia University's Bicentennial celebration, and since the university has long been engaged in enterprises of public moment, it was doubly fitting that recognition be given to the activity that enlarges the world's "access to knowledge." Accordingly, the Trustees of the University and the Directors of its Press decided to signalize the 200th year of Columbia's existence by publishing some samples of current scholarship. A full representation was impossible: limitations of time and space exercised an arbitrary choice. Yet the Bicentennial Editions and Studies, of which the titles are listed on a neighboring page, disclose the variety of products that come into being on the campus of a large university within a chosen year. From papyrology to the determination of molecular weights, and from the state's industrial relations to the study of an artist's or poet's work in its progress toward perfection, scholarship exemplifies the meaning of free activity, and seeks no other justification than the value of its fruits.

JACQUES BARZUN

PREFACE

IT WAS JUST ABOUT FIFTY YEARS AGO that another Columbia professor wrote a book on the development of the frog egg. That book, by Thomas Hunt Morgan, concerned itself with the then current problems of the relation of the gray crescent to first cleavage, the relation of first cleavage to the bilaterality of the embryo, and the effects of gravity, pressure, and other variables in the development of the egg. Many morphological problems were formulated but few were answered. During the intervening years many of these morphological problems have been solved by researches using vital staining, transplantation, and explantation in the developing frog egg. The concepts of presumptive value, prospective potency, and embryonic induction now are well established.

In writing this book on the energetics of development we wish to suggest physiological problems, many of which have not been solved at the present time. As was the situation for the morphological embryology of the early 1900s, we now have a wealth of experimental results in physiological embryology, but we still lack some unifying concepts that must be formulated before the direct coupling between energetics and differentiation can be comprehended fully. It is our hope that the next fifty years will see the same sort of progress in the energetics of development as occurred in morphological embryology after the publication of Morgan's book.

Our general policy during the writing of this monograph was to restrict the work to *Rana pipiens* in order to avoid species differences. Studies on other species have been included only when such investigations were not available for *Rana pipiens*. In addition, we have for the most part limited the papers reviewed to those appearing after 1949, since Brachet in 1950 reviewed the literature up to that time.

Many, although not nearly enough, investigators working in a large number of laboratories have contributed to the literature of chemical embryology. Among this group Columbia is generously represented. The following investigators whose contributions are cited in this monograph worked in Columbia laboratories: Robert Ballentine, Adolph I. Cohen, Samuel Graff, Philip Grant, John R. Gregg, Eugene Healy, J. Gordin Kaplan, Liselotte Mezger-Freed, Florence Moog, John A. Moore, Norma Ornstein, and L. C. Sze.

L. G. BARTH
LUCENA J. BARTH

Mader's Cove, Nova Scotia
September 9, 1953

CONTENTS

General Editor's Preface   v

Preface   vii

One: Introduction   3

The aim of studies on the energetics of development   3
A qualitative description of energetics   4
The energetics of several phases of differentiation   6
   Chemodifferentiation in the determination of cell types   7
      The coupling of energy with the synthesis of large molecules
—The precursor hypothesis of chemosynthesis—Release mechanisms in embryonic determination.
   Reduplication after determination of cell types   9
   Structural differentiation and energetic coupling   10
Specific problems in the energetics of the embryo   10
   The time of transfer of energy to development   10
   Distinction between resting and developmental metabolisms   11
      Special difficulties of distinction during development—The use of embryonic induction to solve the problem.
   Growth metabolism vs. the metabolism of differentiation   12
   Metabolism of determination vs. differentiation metabolism   12

Two: The Storage of Energy   14

Visible changes during growth of the frog oocyte   14
Biochemical synthesis during growth of the frog oocyte   16

Incorporation of phosphate during growth of the oocyte — 18
    Rates of synthesis of P-containing compounds—The site of phosphorylation—Differences between phosphate fractions in the incorporation of P^{32}.
    The effect of the physical state of oocyte components upon incorporation P^{32}—The effect of surface-volume ratios upon incorporation of P^{32}.

Three: The Release of Energy — 25

Difficulties in the determination of the pattern of utilization of stored reserves — 25
  General — 25
  Difficulties encountered in determining utilization — 25
    Insignificance of amounts used vs. amounts stored—Variability between eggs from different females—Conditions for significant analyses.
Respiration of the frog egg during development — 27
  General shape of the normal curve — 28
  Respiration of "blocked" gastrulae — 27
  The relation of respiration to cell division — 29
Carbohydrate metabolism of the frog egg during development — 31
  Glycogen utilization — 31
  Nature of glycolysis in the frog egg — 32
    Lactic acid production and the Pasteur effect—Comparison of respiration and lactic acid production during early development—The reoxidation of lactic acid.
  Coupling between phosphorylation, respiration, and carbohydrate breakdown — 36
  Glycolysis in homogenates of the egg — 39
The effect of inhibitors on development and metabolism — 42
  Special difficulties in the frog egg for inhibitor studies — 42
  Studies with sodium fluoride — 43
    The effect on the respiration of whole eggs—The effect on anaerobic glycolysis of whole eggs—The effect on respiration of explants.
  The effect of sodium monoiodoacetate on development, respiration, and glycolysis — 46
  The use of homogenates for inhibitor studies — 47

The metabolism of blocked hybrid embryos	48
Summary of *pipiens* × *sylvatica* respiratory metabolism—General character of hybrid block—Respiration of *pipiens* × *clamitans* cross.	
Phosphate metabolism during development	50

Four: Localization of the Release of Energy — 53

Special difficulties in the analysis of egg fragments	53
Differentiation of fragments *in situ* vs. in explant	53
Unequal dispersion of cell inclusions	54
Variations in the dissection of fragments: location—size	54
Respiration of the parts of the frog gastrula	55
Facts	56
Interpretation	57
The search for qualitative differences in the respiration of the organizer region	59
Inhibitor studies on the respiration of parts of the gastrula	59
Anaerobic ammonia and CO_2 production by fragments	62
Interpretation	63
Heterogeneity in the distribution of five variables within the gastrula	64
Investigations of local differences within the gastrulae of other species	66
Extractable fraction, fat, total carbohydrate, dipeptidase, and glycerophosphatase activities and oxygen consumption of *Ambystoma mexicanum*	67
ATP and ATPase activity in *Bufo vulgaris*	67
The metabolism of the dorsal lip during induction	68
Glycogen utilization	68
Oxygen consumption	69
Criticism of the method	69

Five: Protein Metabolism — 70

CHANGES DURING DEVELOPMENT	70
Nitrogen metabolism during development	71
Reactivity of frog egg proteins	73
CHEMICAL AND ELECTROPHORETIC STUDIES ON FROG EGG PROTEINS	74

Chemical properties of the KCl extract 75
 Review of properties of R and S reported in previous papers—Experiments exposing further properties of S—Binding of ATP to some component of S—Metal catalysis of the phosphate deficit—Effect of $HgCl_2$ and $MnCl_2$ on phosphate liberation in the presence and absence of ATP—The properties of acid-treated S.

Fractionation of acid-treated S and incubation tests with enzyme ± ATP 84

Electrophoretic studies 85
 Electrophoretic pattern of the KCl extract and identification of two major components 85
  Whole S at stages 8 through 25 + —The patterns of S_1 and of S_2.
 The correlation of electrophoretic pattern with the behavior of components during incubation 88
  Patterns of "acceptor" and of dephosphorylated phosphoprotein—Patterns of S alone and of S + R recovered after incubation—Patterns of incubation mixtures paired with incubation data on aliquots.

The direct determination of alkali-labile phosphate after incubation with and without ATP 94
 Explanation of phosphate deficit in presence of ATP 96
 Source of the surplus phosphate with acid-treated S and ATP 96

Analysis of the effect of denaturation of S and of the role of ATP 97
 Comparison of effects of heat- and acid-denaturation 98
 The effect of concentration of ATP on the liberation of phosphate from phosphoprotein, normal and denatured 99
 Separation of S_1 and S_2 on the basis of acetate solubility 99
  Incubation tests on fractions—Interaction between fractions.
 The effect of S_1 and of ATP on the solubility of phosphoprotein 100
 The roles of S_1 and S_2 in the ATP-denaturation effect 103

Discussion 105
 The potential role of ATP in phosphoprotein breakdown 105
  Indirect control of amount and locus of breakdown—Importance of the new protein, S_1.

Derivation of the proteins from yolk 106
    Use of Panijel's method for yolk extraction—Comparison with previous work on yolk proteins.
  Speculations on the role *in vivo* of ATP in yolk utilization by the developing embryo 107
    Conditions in egg controlling rate of phosphate liberation—Correlation of protein and phosphate metabolisms—Effect of ATP on denatured cell proteins.
Methods 110
  General remarks 110
  Conversion of raw data 110
  Basic procedures for two kinds of experiments 110
    Procedure for obtaining components of the KCl extract and for determining enzymatic liberation of P by various combinations of these components 111
    Procedure for electrophoresis of S proteins 112

Literature Cited 115

Acknowledgments 118

FIGURES

1. The Growth Cycle of the Oocyte of *Rana temporaria* — 15
2. The Increase in the Rate of Oxygen Consumption with Increase in Time of Development — 28
3. The relation of Cell Number to Rate of Respiration during Development — 30
4. The Oxygen Consumption of Frogs' Eggs in Air after Exposure to Low Oxygen Tension — 35
5. The Formation of Lactic Acid from Glycogen by the Action of the Enzymes in Gastrular Homogenates — 40
6. The Relationship between Lactic Acid and Bound Carbon Dioxide during Anaerobiosis — 44
7. The Method of Dissection and the Resultant Fragments of the Gastrula — 56
8. The Method of Dissection of the Gastrula for Experimental Analysis — 60
9. Electrophoretic Diagrams of Yolk Proteins at Different Stages of Development — 86
10. Electrophoretic Diagrams of Yolk Proteins in KCl Extract — 87
11. Electrophoretic Diagrams of S Proteins Recovered after 30 Minutes Incubation — 90
12. Electrophoretic Diagrams of S Proteins from Two Incubation Mixtures — 91
13. Electrophoretic Diagrams of S Proteins Recovered from Incubation Mixtures — 92
14. Electrophoretic Patterns of S Proteins Recovered from Incubation Mixtures — 93

15. Electrophoretic Diagrams of S Proteins Recovered from Incubation Mixtures   94
16. The Effect of Concentration of ATP on Phosphoprotein Splitting in Normal and Acid-treated S   101
17. The Effect of Concentration of ATP on the Splitting of Native Phosphoprotein Isolated from Its Complex with S_1 105

TABLES

1.	The Phosphate Fractions of Washed Yolk and Whole Egg	17
2.	The Relative Increase in Phosphate Compounds during Growth of Oocyte	18
3.	The Distribution of P^{32} in Various Classes of Oocytes	20
4.	The Relative Specific Activity of Phosphate Fractions during Oocyte Growth	23
5.	Respiration and Anaerobic Lactic Acid Production	32
6.	The Reoxidation of Lactic Acid in the Frog Gastrula	33
7.	The Decrease in Labile Phosphate During Anaerobiosis	37
8.	Dephosphorylation and Rephosphorylation of Labile Phosphates in the Frog Egg	38
9.	Lactic Acid Production from Glycogen in Homogenates of Eggs at Various Stages of Development	41
10.	The Distribution of Phosphorus in Phosphate Fractions of the Developing Egg	51
11.	The Rates of Respiration of Fragments of the Frog Gastrula	57
12.	The Rates of Respiration of Fragments of the Frog Gastrula Expressed in Various Terms	58
13.	The Effect of Various Chemical Compounds on Respiration of Fragments of the Gastrula	61
14.	The Distribution of Certain Variables in the Gastrula	64
15.	Balance Sheet for Nitrogen Compounds in the Egg	71
16.	Binding of ATP to a Yolk Protein, S	76
17.	Bound ATP and Phosphate Deficit	77
18.	The Effect of Metallic Ions on Phosphoprotein Splitting	78
19.	The Effect of Mercury and Manganese Ions on Phosphate Liberation	80

20. The Effect of ATP and Manganese on Phosphate Liberation from Acid-treated Phosphoprotein	82
21. Properties of Fractions of Acid-treated S	85
22. Properties of Fractions of S used for Electrophoresis	88
23. The Effect of ATP on Enzymatic Splitting of Normal Phosphoprotein	95
24. The Effect of ATP upon Enzymatic Liberation of P from Acid-treated Phosphoprotein	97
25. The Effect of Heat Denaturation of S on Enzymatic Splitting of Phosphoprotein	98
26. The Effect of Concentration of ATP on Phosphoprotein Splitting in Normal and Acid-treated Preparations of S	99
27. The Phosphoprotein Content of Two Fractions of S	102
28. Solubility of Fractions of S at Different Hydrogen Ion Concentrations	103
29. The Effect of ATP on Splitting of Fractions of Yolk Proteins	104

THE ENERGETICS OF
DEVELOPMENT

one

INTRODUCTION

THE ULTIMATE GOAL of studies of the energetics of development is to explain the various aspects of development in terms of chemical reactions. If the chemical reactions which occur prior to differentiation of cells are known, then it will be possible to control the differentiation of cells. The importance of studies on the energetics of development lies not so much in the mere knowledge of how much energy is necessary, but rather in how the energy is transformed during development to produce growth and cellular differentiation. Consequently we are more interested in the chemical reactions which furnish, transfer, and accept energy than in precise heat measurements during development. Knowledge of the compounds used for combustion and their potential energy is only a beginning toward a more significant study of the paths by which the energy of oxidations is canalized into various morphogenetic processes. Yet it is important to establish the source of energy, for this marks the beginning of a trail which may be followed, however devious the pathway, to the final use of the energy to complete a morphological process.

The embryologist is fortunate in having a plan or pattern provided by the muscle physiologists and biochemists. The early studies on the heat production of muscle contraction and the utilization of glycogen did little to clarify the process of contraction. The studies did however provide the starting point for a long, painstaking series of researches into the chemical reactions subsequent to the disappearance of glycogen. The final arrival at the acceptor for energy in the proteins of the muscle fibers was made only after the discovery of an energy transfer system. Thus the distinction between storage of energy, release of energy, transfer of energy, and utilization of energy by means of a chain of coupled chemical reactions became a pattern for the study of other physiological processes. The embryologist therefore realizes that the establishment of certain compounds as sources of energy is merely the beginning of an extended series of researches on the chemical transformations of these substances and the chemical reactions which are coupled with this chemical transformation.

A QUALITATIVE DESCRIPTION OF ENERGETICS

Let us begin then by establishing a general pattern of the energetics of any process. In the first place, the energy for any physiological or developmental process comes from the free energy of a chemical reaction. A chemical compound, said to be energy rich, is transformed to another form termed energy poor. The type of reaction is exergonic, and we speak of a loss in free energy during transformation. The energy-rich compound is the source of energy. The release of energy may occur by a variety of chemical reactions. A compound may be oxidized with a concurrent decrease in free energy; it may be hydrolyzed; inorganic phosphate may split off a compound; or the acetyl group may be separated from a larger molecule. In each case the resulting components of the chemical reaction are relatively energy poor. Conversely, to form an energy-rich compound from its energy-poor components, free energy must be supplied from some outside source.

This outside source must be another energy-rich compound which when transformed or split chemically releases free energy. Thus the decrease in free energy occurring during one chemical reaction may be taken up by a second reaction which requires free energy. This second reaction is characterized as endergonic and during the reaction a gain in free energy occurs. In order that one exergonic reaction donate its free energy to another reaction (endergonic) there must be some sort of coupling between the two reactions. Such a coupling is termed energetic coupling. In general, the coupling is by means of a common

INTRODUCTION

radical, such as the phosphate radical, or by means of an oxidative reaction involving enzymes and coenzymes. In this way the concept of an energy donator transferring its energy to an energy acceptor arose. When the phosphate bond is involved we speak of phosphate being transferred from a phosphate donator to a phosphate acceptor. Compounds may form a chain along which phosphate is transferred, and with the transfer of the phosphate a transfer of energy may be achieved. Thus the phosphate radical of phosphocreatine may be donated to adenosine diphosphate to form adenosine triphosphate. The adenosine triphosphate then may donate its phosphate to glucose-6-phosphate to form hexose diphosphate, which then may be split and the resultant products oxidized. The final transfer of phosphate takes place between adenosine triphosphate and muscle proteins. In this way the energy for muscular contraction is obtained from the changes in free energy of the phosphate bond.

The change in free energy during a chemical reaction is a function of the equilibrium constant of the reaction. If the reaction goes almost to completion then the decrease in free energy is great and the reaction becomes, potentially at least, a source of high energy for transfer to reactions which do not proceed forward to any appreciable extent. The change in free energy thus is a measure of the driving force of a reaction, that is, the tendency of the reaction to proceed in the forward direction. A large decrease in free energy during the splitting of a compound also indicates that the compound is unstable. A small decrease in free energy of a reaction means that the compound is relatively stable. Of course for a reaction to proceed, the correct conditions of temperature, hydrogen ion concentration, and catalysts must be present. Otherwise, stable or unstable, a reaction will not reach equilibrium in any reasonable time.

As a concrete example we may use adenosine triphosphate, a compound in which the terminal phosphate radical is bound to the rest of the molecule by an energy-rich bond. In the presence of an enzyme, adenosine triphosphatase, phosphate is released from adenosine triphosphate and thus adenosine diphosphate and phosphoric acid are formed. When equilibrium is reached practically all the adenosine triphosphate is gone and we have large amounts of adenosine diphosphate and phosphoric acid in solution. There is a large decrease in free energy during the reaction and adenosine triphosphate may be regarded as unstable. The splitting of the compound to adenosine diphosphate and phosphoric acid has occurred spontaneously without the addition of free energy, since we have merely introduced a catalyst which does not affect the equilibrium but only hastens its attainment.

If we were to start with high concentrations of adenosine diphosphate and phosphoric acid and the enzyme and allow the two components to combine to form adenosine triphosphate, only traces of the latter would be formed. The reaction of adenosine diphosphate with phosphoric acid to form adenosine triphosphate is an endergonic reaction and requires an external source of free energy. This external source may be obtained from an exergonic reaction which yields free energy. If phosphocreatine splits to give creatine and phosphoric acid, the reaction proceeds nearly to completion with a large decrease in free energy. Therefore the energy of this reaction can be used to synthesize adenosine triphosphate from adenosine diphosphate. Phosphocreatine will donate its high energy phosphate to adenosine diphosphate, and adenosine triphosphate is formed. There are other means of forming adenosine triphosphate from the diphosphate and in each case the synthesis involves an energy-rich donator of phosphate.

In general, most of the larger molecules in protoplasm are energy-rich in the sense that when split into smaller molecules enzymatically they break down almost to completion with a large decrease in free energy. Conversely, in order to synthesize these large molecules from smaller ones a large amount of free energy must be donated at some step in synthesis by another chemical reaction. This free energy for synthesis ultimately comes from the oxidation of various substrates.

THE ENERGETICS OF SEVERAL PHASES OF DIFFERENTIATION

The embryologist is confronted with the task of tracing the chain of reactions from the stored energy-rich compounds, glycogen, phosphoproteins and proteins, phospholipids and lipids, to the final acceptors of energy within the differentiating cells. In contrast to the large amounts of free energy needed for muscular contraction, that required for differentiation of an embryonic cell into a specific cell type may be very small. It may be difficult to distinguish this small amount of energy when larger amounts are being diverted into maintaining the cell structure, into the process of cell division, into the process of synthesis of nonspecific protoplasm, into the synthesis of genes, and into the process of cell movement.

Experimental embryology has given us the times at which various cells become irreversibly determined to develop into various cell types. This first step in differentiation is termed chemodifferentiation, and we have as yet no clear evidence that energy is needed at this stage of the process of differentiation. Thus we can merely speculate on possibilities.

If a large molecule must be synthesized from smaller fragments for the determination of a cell type, then it is clear that the process of determination must be coupled with a reaction exhibiting a loss in free energy. If the exergonic reaction is coupled with the synthesis of the large molecule then energy is transferred. Such a relationship would make it possible to trace the chain of energy transfer to the final acceptor very much as has been done in the study of muscle metabolism.

For example, it is a reasonably well established fact now that glycogen is utilized as an energy source during gastrulation and that glycogen is used more rapidly in the organizer region of the amphibian gastrula. During gastrulation the neural plate becomes determined and will no longer differentiate into other structures. If the energy available from the oxidation of glycogen is necessary for the determination of the neural plate, then a study of the chain of reactions from glycogen to carbon dioxide and water should lead to a point where some reaction furnishes the free energy required for the synthesis of a compound necessary for this determination. Anywhere along the chain of reaction the decrease in free energy of a reaction may be transferred to another reaction and an energy-rich compound may result. This energy-rich compound may then be utilized to synthesize a large molecule from smaller fragments.

Another possible relationship between the formation of a large molecule and the process of determination must be considered. A precursor of the large molecule may be present in the cell and by a reaction needing little or no external energy source the precursor may be converted into the large molecule. Thus it has been suggested that the specific proteins of cells are not synthesized individually from amino acids with the obvious need for large amounts of free energy for the formation of the peptide bond. Rather this explanation is offered: that a general precursor protein, a proteinogen, is first synthesized at the expense of the energy of respiration. This proteinogen present in all cells is then converted to the specific proteins of various cell types by an exergonic reaction, i.e., one which needs no energy and may even result in a decrease in free energy. With this hypothesis of the formation of specific proteins a coupling of the formation of each individual specific protein with a specific energy-producing reaction is not necessary. A general exergonic reaction produces the energy for the formation of the proteinogen which in turn is converted into the various specific proteins without further addition of energy.

If this mechanism operates to determine embryonic cells to form specific cell types, then there is no direct relationship between energy-producing reactions such as oxidations and embryonic determination. A study of the chain of reactions providing energy would simply lead to the coupling with reactions forming a proteinogen and we would still be left with the main problem of the conversion of the proteinogen into specific proteins. In this schema energy-producing reactions would have a minor role and the chemical embryologist might better begin with a study of the proteins of the egg rather than its respiration and metabolism.

Finally, let us turn to release mechanisms as a formal explanation of the determination of embryonic cells. Here again we might assume that large specific molecules are necessary for determination, but this time they preexist in the egg protoplasm in a bound, inactive state. When a cell becomes determined, say, to form a neural plate cell, a specific protein is released from a bound form. Each cell then would contain all the specific proteins for all the specific cell types and the nature of the differentiation of any cell would depend on which specific protein was released. The release of specific proteins would occur by the action of specific inductors or by a specific state of the cytoplasm at the time of determination. In any event no energy would be necessary for the release of the protein and the reaction might even be a source of energy.

Such a view of embryonic determination shifts the whole problem of the energetics of development back to the developing oocyte. In the ovary the energy of respiration would be transferred to reactions leading to the synthesis of the specific proteins of various cell types. We would then need to study the mechanism of storage of such elements as yolk and to investigate the nature of the cytoplasmic proteins. The specific proteins of various cell types would be present in the unfertilized egg in the form of inactive proteins and development would become a mechanism for the orderly activation of these proteins. There is no doubt that protein synthesis occurs during the growth of the oocyte of the amphibian egg, but as yet there is no evidence that these proteins are anything more than raw materials to be used in the synthesis of protoplasm during growth of the embryo.

The foregoing discussion of the nature of the chemical changes at the time of determination of various cell types restricts the problem to the very first step in differentiation. After determination the cell may continue to divide producing daughter cells which are likewise determined. This means that the chemical change at the time of determination is propagated, that some reduplicating unit is formed. This conclusion follows from the fact of self-differentiation of cell types in explants. Here all inductors have been removed and thus the initial stimulus for synthesis or for release of specific proteins, as the case may be, is no longer present. Yet the cells resulting from division are determined and thus the specific protein reduplicates itself in the absence of inductors.

This phase of reduplication may or may not require energy. If a large supply of proteinogen is present at the time of determination, specific proteins could reduplicate themselves with little or no external energy source. On the other hand, if the specific proteins are formed from smaller fragments, energetic coupling must occur.

After embryonic determination and after a period of cell division an active structural differentiation of the determined cell type takes place. Changes in form and chemical constitution occur and in many instances there is extensive cell migration, as with neuroblasts and myoblasts. In connective tissue active secretion of fibrils and matrix requires both a source of substrates and an energy source. The phase of structural or histological differentiation probably is coupled energetically with respiration and glycolysis, since by this time the stored materials probably are exhausted and synthesis depends upon small molecules obtained from blood.

The problem of the energetics of the final phase of differentiation, functional differentiation, merges with the problem of the energetics of function in the adult. Almost certainly every cellular function is coupled with an energy source.

Undoubtedly the embryologist investigating the energetics of development is faced with a problem that is both complex because of the many variables discussed in the preceding pages and difficult because of the small amounts of material available for chemical analysis. The difficulty of the problem fortunately has been lessened by the emergence of ultramicrochemical analysis and ultramicrophysical instrumentation.

SPECIFIC PROBLEMS IN THE ENERGETICS OF THE EMBRYO

The complexity of the problem remains. On the part of energetics of any process we must distinguish between energy storage, donation or release of energy, transfer of energy, and utilization or acceptance of energy. Along this chain we must be on the alert for energetic coupling at any step in the release of energy. The developmental process is even more complex and certainly less well known from the standpoint of chemical reactions. The acceptors of energy are many and, as we have seen, the time at which energy is needed is not known. It is conceivable that all the energy needed is transferred from respiration before fertilization. On the other hand, the energy of respiration may be utilized continuously with development.

In any case for the purpose of any analysis we distinguish a resting metabolism, i.e., the energy for maintenance, and an active metabolism, the energy for development. The resting metabolism presents a difficulty peculiar to the process of development; namely, development is irreversible. Whereas in the study of muscle metabolism a muscle may begin in a resting state, be stimulated to contract, undergo relaxation, and return to a resting state, no such cycle is possible in development. Development is a continuous, irreversible process, and any stage we choose is simply a resting stage for the next step in the process. We cannot measure the respiration during the formation of the neural plate and then refer it to a resting metabolism during gastrulation because during gastrulation we measure the resting metabolism of the gastrula plus the metabolism of the gastrulation process. If we had some satisfactory means of stopping development and stimulating it once more, a true resting metabolism for any process might be obtained. The attempts to stop development by inhibitors or by hybridization, however, usually result in further complications depending upon the nature of the block to development.

Possibly the best we can do at this time is to take advantage of embryonic induction to distinguish between the metabolism of development as opposed to a resting metabolism. If two equal portions of the presumptive neural plate are selected and one is induced by the presumptive notochord to form a neural plate while the other is not so induced, the difference in metabolism between the two preparations may be a measure of the metabolism of the process of embryonic determination of the neural plate. The technical difficulties involved make the procedure less clearcut than as stated. The metabolism of the explant may be altered drastically by the operation and that of the inductor as well. These abnormal metabolisms well might distort any differences in metabolism in the two preparations.

Once having obtained the difference between resting metabolism and developmental metabolism, there still remains the distinction between growth metabolism and the metabolism of differentiation. When cell division and synthesis of protoplasm proceed along with differentiation it is difficult to separate the two processes. If one uses the presence or absence of the inductor to determine the developmental metabolism he can, of course, assume that the processes of growth and cell division occur both with and without the inductor, and that he then is measuring the metabolism of differentiation.

Finally, if success is attained in distinguishing the metabolism of differentiation from growth and resting metabolism there is still the matter of discriminating between the metabolism of embryonic determination and that associated with cell movements and changes in cell shape during histological differentiation. Here the time tables for embryonic determination supplied by the experimental embryologist will be of great value. The methods of measuring metabolism will have to be refined so that any change must be detected at the time of embryonic determination and before histological differentiation commences. For example, in the usual combination of the inductor of the neural plate, the dorsal lip, with the presumptive neural plate the two explants are combined and allowed to develop for two to three days. By this time a neural tube has formed and possibly sensory structures also have formed. A change in metabolism measured over this period of two or three days would be difficult to interpret. The difference in metabolism might be associated with embryonic determination or with the mechanics of cell elongation and the folding of a plate to form a tube. If the energetics of embryonic determination be the problem, then metabolic measurements must be made at the time when the presumptive neural plate is irreversibly determined but not histologically differentiated. This time would be at the end of gastrulation in the intact gastrula.

The problem of the energetics of development is presented in this chapter in its most complex aspect and as such appears to be very difficult of solution. The elucidation of the many reactions involved in the storage, release, transfer, and utilization of energy alone would be a very ambitious project. When we add to this problem the separation of development into its many processes the difficulties multiply. Such complexities and difficulties, however, have been encountered in the energetics of muscular contraction and the complexities have to a large extent been simplified and the experimental difficulties overcome. The embryologist may anticipate that with further fact finding in the field of energetics of development, some direct coupling between energy production and development will emerge.

two

THE STORAGE OF ENERGY

OUR PROBLEM of the energetics of development begins chronologically with the storage of energy-rich compounds in the developing frog oocyte. During the development of the egg after fertilization these compounds undergo a process of gradual breakdown into smaller fragments with an overall decrease in free energy. Thus a cycle of storage and release of energy begins with the germ cell in which glycogen, phospholipids, lipids, phosphoproteins, and proteins accumulate during its growth period in the ovary. This process of storage results in a mature, unfertilized egg with large amounts of these energy-rich compounds and a relatively small amount of cytoplasm and nucleoproteins. Between fertilization and the tadpole, stage 25, the stored compounds are broken down and cytoplasm and nucleoproteins again predominate. The cycle ends with the segregation of the germ cells in the developing ovary. During the period from fertilization to stage 25 little is added to the system save oxygen and water, and the dry weight of the egg decreases through loss of excretions.

VISIBLE CHANGES DURING GROWTH OF THE FROG OOCYTE

Let us begin by examining the gross visible changes which occur during the growth of the oocyte of the frog. This period takes three years and may be contrasted with the rapid utilization of the stored compounds in about ten days. The small oocytes, about 0.05 mm. in diameter, grow as shown in Figure 1 (Grant, 1953) to a mature oocyte about 1.5 mm. in diameter, an increase of about 27,000-fold. Most of this growth takes place during the three summers, as is indicated in the graphs by a flattening of the growth curve during the winter months.

At any definite time, say October of the third year (Figure 1), three groups of oocytes are present. The largest, about 1.5 mm. in diameter, group 1, are those eggs which will normally be

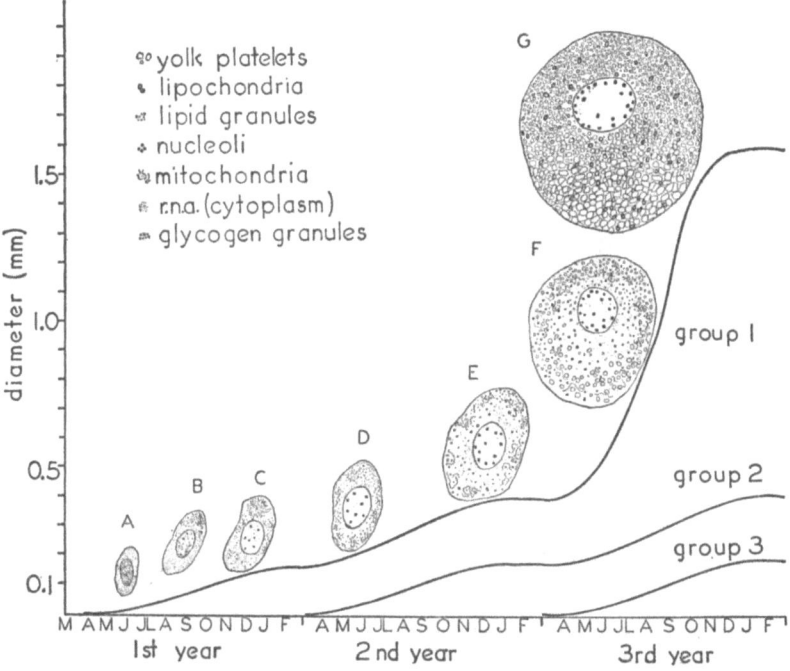

FIGURE 1. THE GROWTH CYCLE OF THE OOCYTE OF THE FROG *Rana temporaria*
The histological structure of the oocyte is shown in semidiagrammatic form. From Grant, 1953.

released in the following March. Other oocytes, group 2, about 0.4 mm. in diameter, have grown for two summers and require one more for maturity. Group 3, finally, represents a single summer's growth to about 0.1 mm. in diameter. These three groups of eggs coexist in the ovary and form a continuous series from the smallest oocyte to the mature egg.

The microscopic structure of the oocyte is shown in semidiagrammatic form in Figure 1. Here the various types of granules present in the oocyte are shown at various intervals in the three-year period. Seven types of granules are depicted, ranging from the large yolk platelets to the small granules of ribosenucleoprotein (r.n.a.). In the nucleus, which becomes a large germinal vesicle in the mature egg, nucleoli are formed.

BIOCHEMICAL SYNTHESIS DURING GROWTH OF THE FROG OOCYTE

According to Panijel (1951) whose monograph on oogenesis should be consulted for details concerning the histological and biochemical descriptions of the growth of the oocyte, two periods of growth may be distinguished. The early period of growth is characterized by the formation of lipid granules and glycogen granules, while the later growth is largely a result of the development of yolk platelets containing protein.

The very early oocytes somewhat less than 0.05 mm. in diameter contain a relatively large, dense nucleus and a homogeneous cytoplasm with ribonucleic acid granules. The lipid granules appear in oocytes of about 0.05 mm. in diameter, while glycogen granules form somewhat later (in oocytes 0.2 mm. in diameter). Finally, the yolk platelets develop when the oocyte has attained a diameter of 0.35 mm., and the platelets grow in size and increase in number until the oocyte is mature (about 1.5 mm. in diameter). The sequence in the appearance of the three classes of compounds may be compared with the times at which they are utilized by the developing eggs. As nearly as chemical measurements can determine, the glycogen begins to disappear during gastrulation, the lipids just before hatching, and the yolk platelets at about stage 22, about two days after hatching. This statement must not be taken to mean that small, unmeasurable amounts of all three classes of compounds are not used in early development. Because of the large amounts of glycogen, lipids, and yolk present in the fertilized egg in relation to the amount of cytoplasm present, it is not possible to detect small losses in any of these three fractions.

Some idea of the high concentrations of the three classes of compounds is indicated by measurements of glycogen, which in the young oocyte is present as 0.9 percent of the dry weight of the oocyte but in the mature egg reaches a concentration of 8.1 percent (Panijel, 1951). Lipids are stored until they reach a concentration of 25 percent of the dry weight of the egg, while the yolk forms an even greater percentage of the dry weight of the egg.

The yolk, although mostly protein, is not a pure chemical compound but contains other substances (Table 1). This table deals only with the phosphorus-containing compounds and is not a complete analysis of the yolk platelet. In addition to the phosphoprotein there is an even larger amount of another protein

TABLE 1

THE PHOSPHATE FRACTIONS OF WASHED YOLK GRANULES COMPARED WITH THE SAME FRACTIONS IN THE ENTIRE EGG

The values are in micrograms per egg. From Panijel, 1950, as summarized by Grant, 1953.

Fraction	Whole Egg	Yolk	$\frac{Yolk}{Whole\ Egg} \times 100$
Phosphoprotein P	6.31	5.79	92
Nucleic acid P	2.17	0.74	34
Lipid P	2.33	1.23	53
Acid-soluble P	0.74	0.30	41
Inorganic P	0.12	0.19	—
Organic acid-soluble P	0.36	0.11	31
Total P	11.74	8.20	70

and very small amounts of enzyme. Table 1 shows, first of all, the complexity of the yolk platelet, containing as it does phosphoprotein, ribosenucleic acid, phospholipids, organic acid-soluble phosphates, such as adenosine triphosphate, and some inorganic phosphate. The smaller molecules must be in chemical combination with the larger, insoluble molecules of the yolk since prolonged washing does not remove inorganic and acid-soluble phosphates. The table also shows that the yolk contains about 70 percent of the phosphate of the entire egg, and that in the case of the phosphoprotein over 90 percent of it is localized in the yolk platelets.

Panijel (1950, 1951) has separated the yolk platelets into two fractions. One fraction is composed of large platelets about 15 microns in length and comprises 90 percent of the total. The other fraction contains small yolk platelets about one micron in length. These two fractions differ in chemical constitution and enzymatic activity. The small platelets contain relatively more ribosenucleic acid and more of the enzyme phosphoprotein phosphatase but much less lipid than the large platelets.

INCORPORATION OF PHOSPHATE DURING GROWTH OF THE OOCYTE

In three years the growing oocyte acquires or builds up a large reserve of glycogen, lipid, and protein. During development the glycogen and lipid are largely oxidized but the protein oxidation is negligible. The protein stores, therefore, are chiefly a reserve for transformation into cytoplasmic proteins. The phosphoprotein in addition is a reserve of inorganic phosphate, which is necessary for phosphorylation during oxidation and for synthesis of nucleic acids. Because of the prominent role of phosphate in cellular metabolism a study of its incorporation into the growing oocyte has been made (Grant, 1953). Five stages in the growth of the oocyte were selected for analysis and each stage was analyzed for various phosphate-containing compounds. The results show the relative growth in amount of various phosphate fractions of the developing oocyte (Table 2). The overall increase in total phosphate in the oocyte is from .0487 micrograms in the smallest oocyte to 10.36 micrograms in the mature oocyte, or a 212-fold increase.

TABLE 2

THE RELATIVE INCREASE IN VARIOUS PHOSPHATE-CONTAINING COMPOUNDS DURING THE COURSE OF GROWTH OF THE OOCYTE

The values are in terms of micrograms of P in 100 oocytes. Class A = 0.05 to 0.25 mm. oocytes; class B = 0.05 to 0.40 mm. oocytes; class C = 0.50 to 0.90 mm. oocytes; class D = 0.90 to 1.00 mm. oocytes; class E = 0.90 to 1.50 mm. oocytes. Class D values are taken from one experiment only. From Grant, 1953.

Fraction	OOCYTE CLASS				
	A	B	C	D	E
Total P	4.87±1.17	11.71±2.92	49.94±13.49	137.30	1036±275
Phosphoprotein P	0.76±0.48	1.87±1.65	3.91± 3.42	23.95	631.3±78.8
Nucleic Acid P	3.24±1.48	5.51±2.24	19.60± 4.49	26.10	217.3±98.2
Phospholipid P	0.24±0.19	0.71±0.61	7.88± 2.26	22.60	232.8± 9.1
Acid-soluble P	1.23±1.14	3.07±1.44	8.25± 2.30	24.13	73.9±41.0

This is considerably less than the increase in volume over the same period, about 1000 times, indicating that the growth of non-phosphate-containing compounds is occurring at a faster rate than that of phosphates. Among the phosphate fractions themselves it is seen that the phospholipid increases most rapidly during early growth, an increase of 33-fold between stage A and stage C, while nucleic acid increases by a factor of 6 and phosphoprotein by a factor of 5 for the same interval. During the second phase of growth, from stage C to stage E, the situation changes. The growth of phosphoprotein becomes much more rapid and there is a 16-fold increase. For this period the nucleic acid increases by a factor of 11, while the phospholipid continues to form at the most rapid rate (about a 30-fold increase).

Phospholipid thus is synthesized by the growing oocyte at a rapid and fairly constant rate, while phosphoprotein first is formed rather slowly and then increases rapidly during the period from stage C to stage E. This period is characterized by the development of the yolk platelets which contain the phosphoprotein.

Nucleic acid increases more rapidly than phosphoprotein during the early period, but less rapidly during the late period. Grant's analysis, in general, agrees with a two-phase growth cycle consisting of a synthesis of lipids and nucleic acid in the early stages and of phosphoprotein in the later stages.

The acid-soluble phosphates, consisting of inorganic phosphate and various organic phosphates, increase by a factor of about 7 from stage A to stage C and in later growth by a factor of 9.

The over-all increases in the various phosphate fractions from stage A to stage E are as follows: (1) phospholipid, $960\times$; (2) phosphoprotein, $830\times$; (3) nucleic acid, $67\times$; (4) acid-soluble, $60\times$. During the same period the overall increase in volume is about $1000\times$. From these figures it is clear that some nonphosphate-containing compound must be synthesized at an even greater rate than are the phosphate-containing compounds in order to account for the volume change. The possibility that the oocyte takes on relatively more water is unlikely since, due in part to the low water content of the yolk, the percentage of water in the mature egg is less than that of an average cell. Studies on the composition of the yolk presented in an ensuing chapter show that the yolk contains a protein present in much larger quantities than the phosphoprotein. It may be that this protein is synthesized even more rapidly than the phosphoprotein and thus accounts for the more rapid increase in volume of the egg as compared with the increase in phosphate-containing compounds.

Returning to the synthesis of the phosphate compounds we may ask whether this synthesis occurs within the growing oocyte or whether it occurs in the tissues of the frog and the compounds

TABLE 3

THE DISTRIBUTION OF P^{32} IN VARIOUS CLASSES OF OOCYTES

The activity of P^{32} is measured as the number of counts per minute for 100 oocytes. This value is divided by the number of counts per minute of the P^{32} injected divided by the weight of the frog in grams. The resultant is divided by 1000.

$$\text{Thus, A/O} = \frac{\text{counts/min./100 oocytes}}{\text{injected act./wt., gms.}} \times 10^{-3}$$

Percent of activity is simply the activity of any P fraction divided by the activity of the total P. Class A = 0.05 to 0.25 mm. oocytes; class B = 0.05 to 0.40 mm. oocytes; class C = 0.50 to 0.90 mm. oocytes; class E = 0.90 to 1.50 mm. oocytes. From Grant, 1953.

	OOCYTE CLASS							
	A		B		C		E	
FRACTION	A/O	% Activity	A/O	% Activity	A/O	% Activity	A/O	% Activity
Total P	1.62		5.49		18.05		31.50	
Phosphoprotein P	0.07	5.49	0.17	3.68	0.45	2.58	10.70	34.60
Nucleic acid P	0.14	11.60	0.33	6.56	0.86	5.25	3.49	16.60
Phospholipid P	0.02	1.67	0.08	1.04	0.21	1.64	0.59	2.14
Acid-soluble P	0.93	80.60	4.60	88.70	16.58	90.60	15.12	46.10

then are transferred to the egg. While a definite answer is not possible in all cases, it seems clear from studies of the incorporation of P^{32} that much of the phosphorylation occurs within the growing oocyte. Injection of P^{32} in the form of inorganic phosphate into the adult female results in the appearance of P^{32} in the oocyte (Table 3, from Grant, 1953). Since the oocyte is exposed to P^{32} for only 24 hours and its growth period is three years, it is highly unlikely that the P^{32} in the egg results from the formation of any additional phosphate compound. Rather the presence of P^{32} shows that the phosphate compounds in the egg form a steady-state equilibrium so that the normal P^{31} forms some sort of equilibrium with the injected

P^{32}. This situation would obtain only if all the enzymes and the energy source for synthesis were present in the oocyte. In the case of acid-soluble phosphates such as adenosine triphosphate it is possible, of course, that the equilibrium between P^{31} and P^{32} occurs in the blood of the female and that an exchange between ATP^{32} of the blood and ATP^{31} of the oocyte takes place. But in the case of the phosphoprotein such a situation is highly unlikely, since the phosphoprotein in the oocyte exists in an insoluble, nondiffusible form in a platelet. The only means by which P^{32} could enter into the phosphoprotein is through the presence of a phosphoprotein phosphatase, an enzyme necessary for the splitting of the phosphoprotein into protein and inorganic phosphate. An exchange between P^{31} and P^{32} then could occur. Further, since the equilibrium of the reaction, phosphate plus protein giving phosphoprotein, is displaced in the oocyte toward a high concentration of phosphoprotein, a constant source of energy must be present. Otherwise the phosphoprotein would be dephosphorylated. The most reasonable assumption in the case of the incorporation of P^{32} into the phosphoprotein fraction of the oocyte is that P^{32} enters either as inorganic phosphate or acid-soluble organic phosphate, and that the exchange between P^{31} and P^{32} occurs within the oocyte itself. A study of Table 3 supports this reasoning. In every class of oocyte analyzed the highest percentage of the P^{32} in the oocyte is found in the acid-soluble P fraction. This in spite of the fact that the acid-soluble P is the smallest fraction of the total P in class E.

Since the activities of P^{32} in Table 3 are expressed in terms of 100 oocytes and the oocytes of different classes differ largely in size, it is only natural that the activity of the total P increases from stage A to stage E. Between these two stages the total amount of phosphate changes from about 5 micrograms to 1000 micrograms. Therefore more phosphate bonds occur where P^{32} can substitute for P^{31}.

Among the various phosphate fractions differences in the incorporation of P^{32} appear. During early growth, stages A through C, where nucleic acids and phospholipids are synthesized, most of the P^{32} (from 80 to 90 percent), is in the acid-soluble fraction. Relatively more P^{32} is in the nucleic acid fraction as compared with the phospholipids, a situation that may result from the greater amounts of nucleic acid P present.

In the mature oocyte, class E, the distribution is different from that in early stages. While the acid-soluble fraction continues to show a high activity, 46 percent, the activity of the phosphoprotein fraction is also very high, 34 percent. This high activity of the phosphoprotein fraction is understandable when we recognize that most of the phosphate in the egg is now in the form of phosphoprotein—60 percent in class E, while in early stages, A to C, phosphoprotein formed only a small fraction of the total, 8 percent, for example, in class C oocytes. Since the percentage of phosphoprotein P increases, therefore, the percentage of activity of P^{32} might be expected to increase, as it does.

In the Class E oocytes the nucleic acid and phospholipid fractions are of interest in that the former contains about 17 percent of the P^{32} and the latter only 2 percent. This difference cannot result from a difference in amount of P^{31} in the two fractions, since Table 2 shows that the phospholipid P is about the same as the nucleic acid P. Possibly the phospholipids, being insoluble, are less reactive than the nucleic acids, of which at least some are in solution and others are in the form of very fine granules. Some of the phospholipids in class E oocytes are located in yolk platelets and compounds in the interior of these platelets would have no chance to exchange their P^{31} for P^{32}.

In considering the incorporation of P^{32} into the components of the oocyte, it is important to consider the degree of dispersion of the components. If a component is in solution, then all molecules are free to react and exchange P^{31} for P^{32}. If the component is in the form of a fine granule the amount of surface where exchange can take place in relation to the volume of the granule is very high. Thus a high percent of activity may be expected, as occurs in the nucleic acid fraction. But if the granule is large such as is the case with the large yolk platelets, then the area of surface in relation to the volume of the granule becomes small and the opportunity for exchange is less than with small granules or for solutions.

When the P^{32} incorporated into various phosphate fractions of the oocyte is expressed in terms of the amount of phosphate present in each fraction the figures in Table 4 are obtained. These values are the activities of Table 3 divided by the absolute amounts of phosphate as given in Table 2. Essentially we arrive at the activity per unit of phosphate present in each fraction. Table 4 shows that for total P the relative specific activity decreases as the amount of phosphate in the oocyte increases. That is, a relatively smaller percent of the P^{32} is exchanged with the P^{31} of the various phosphates. This decrease is understandable

TABLE 4

THE RELATIVE SPECIFIC ACTIVITY OF THE VARIOUS PHOSPHATE FRACTIONS OF THE OOCYTE DURING GROWTH

The values in the table are obtained by dividing the activities of Table 3 by the absolute amount of P of each fraction in Table 2, and multiplying by ten. This gives an activity of each fraction per unit of P of that fraction present in the oocyte. From Grant, 1953.

Fraction	OOCYTE CLASS			
	A	B	C	E
Total P	3.33	4.68	3.62	0.34
Phosphoprotein P	0.92	0.91	1.15	0.17
Nucleic acid P	0.43	0.60	0.44	0.16
Phospholipid P	0.83	1.13	0.27	0.03
Acid-soluble P	7.56	14.99	20.10	2.05

in the later stages of growth of the oocyte where much of the P^{31} is in large granules in which there is no opportunity for exchange of the P^{31} with P^{32}.

Phosphoprotein and nucleic acid P show a decrease in relative specific activity in the mature oocyte, and their values are the same at this stage. This identity means that per mol of P present as nucleic acid P or phosphoprotein P, as much P^{32} will enter nucleic acid as will enter phosphoprotein. The decrease in the relative specific activity of the two fractions in mature oocytes may be in part a result of the fact that these compounds are in the yolk platelets. Ninety-two percent of the phosphoprotein P of the entire egg is present in the yolk, while 34 percent of the nucleic acid P in the egg is found in the yolk platelets. It is doubtful whether very much of the P^{31} of these yolk fractions can exchange with P^{32} unless the fractions are at the surface of the platelet. Similarly the relative specific activity of the phospholipid P decreases in the mature oocyte and this too may result from the localization of 53 percent of this fraction in the yolk platelet. Even an appreciable amount, 41 percent, of the acid-soluble P is found in the yolk platelet, and again, unless it is at the surface, little exchange can occur.

One other factor needs to be considered for the interpretation of Table 4 and that is the decreasing surface-volume ratio of the growing oocyte. With the diameter of the oocyte changing about tenfold from class A to class E, the surface area per unit volume, or more significantly per mol P, decreases rapidly. Since the P^{32} must diffuse from the surface of the oocyte to its center in order to reach all the phosphate fractions, it is clear that the concentration of P^{32} in the center of the mature egg may be relatively low. This situation would affect the rate of exchange of P^{31} with P^{32}. The decrease in relative specific activity with increase in size of the oocyte noted for all fractions in Table 4 may be then a function of a decrease in surface area per unit volume. Two factors thus must be recognized: the decrease in the availability of the phosphate fraction for exchange with P^{32} and the possible decrease in the penetration of P^{32} in the larger, older oocytes as a result of a relative decrease in surface area.

In summary, we find a storage of glycogen amounting to 8 percent of the dry weight of the mature egg and an active phosphate metabolism resulting in the storage of phosphoprotein, phospholipid, and nucleic acid phosphates. We pass on now to the release of energy during development, and from the above facts we might anticipate an active glycolysis coupled with phosphorylation. All the elements for energy transfer through phosphate bonds appear to be present.

three

THE RELEASE OF ENERGY

AFTER FERTILIZATION of the frog egg the compounds synthesized during the growth of the oocyte are utilized after a pattern which as yet has not been determined satisfactorily. We do know that at the larval stage, when the animal begins to feed, all the stored reserves have disappeared. Some have been oxidized, while others have been transformed. In general the glycogen and lipids are oxidized, while the proteins are transformed. The times at which the metabolisms of various compounds begin and the rate of their metabolisms continues to be a fruitful study (Løvtrup, 1953).

DIFFICULTIES IN THE DETERMINATION OF THE PATTERN
OF UTILIZATION OF STORED RESERVES

Before proceeding with our analysis of the release of energy a general consideration of the problem will clarify some of the results obtained. First of all, the frog egg weighs about 1.5 milligrams, and as we have seen in the preceding chapter the stored fat, glycogen, and protein make up most of this dry weight. The active, metabolizing protoplasm is very small in comparison with the stored materials. The demands of the egg for stored energy therefore are slight during the early stages—cleavage, blastulation, early gastrulation. It is practically impossible, therefore, with current physical and chemical methods to detect a decrease in the amount of either fat, carbohydrate, or protein during early development. There is simply too much of these components present to determine whether or not a decrease occurs with development. Thus when it is found by analysis that a significant decrease in the amount of glycogen occurs between fertilization and the end of gastrulation we cannot conclude that the metabolism of glycogen begins during gastrulation. All that can be concluded is that glycogen metabolism begins some time prior to stage 12. It seems to us that it would be fairer to extrapolate a smooth curve from stage 12 back to stage 1, with the implication that glycogen metabolism is going on continuously, rather than to conclude arbitrarily that glycogen metabolism begins when it becomes technically possible to measure a decrease in glycogen content. Similarly, with 25 percent of the dry weight of the egg occupied by lipids, the conclusion that lipid metabolism begins just prior to hatching is not justified, since it is apparent that we could not hope to detect decreases in lipid in cleavage stages. The situation is the same in the case of proteins.

A second consideration may clarify some of the variability of the results of experiments reported. While the dry weight of the frog egg from a particular batch of eggs may be 1.5 mg., there is considerable variability in the dry weight and chemical composition of the eggs from different females. In a series of experiments with eggs from different females the dry weight varied from 1.47 to 1.86 mg., a difference of 0.39 mg. Since the entire loss in dry weight from fertilization to hatching is only 0.2 mg., it is clear that no data of significance can be obtained by measuring the dry weight of one batch of eggs at fertilization and the dry weight of another batch at hatching. The same situation holds for the determination of decreases in glycogen, lipid, and protein.

Some investigators have corrected for differences in dry weight of eggs of two determinations by relating their value for any variable to the dry weight of the egg. This procedure involves an assumption that the chemical compositions of eggs from different females are the same. In our own experience this assumption is not justified. In a lengthy series of measurements of oxygen consumption of frogs' eggs a record of the dry weight of each batch of eggs was made. When the rate of oxygen consumption per unit dry weight of egg was calculated, the rate of oxygen consumption of the larger eggs was higher than that of the smaller eggs (18.4 as compared with 15.0). This means that the chemical composition of the larger eggs is different from that of the smaller eggs with respect to the compounds involved in respiration.

In view of these facts certain conditions for analyses of frogs' eggs may be outlined. In any analysis designed to reveal differences in any variable during development, aliquots of the eggs of a single female must be used for each determination. If the number of eggs from a single female is not sufficient, then several clutches may be combined using equal numbers from each clutch for each analysis.

Even under these conditions a source of error enters. One hundred percent normal development rarely if ever is obtained. Some eggs develop abnormally and some eggs die. The eggs or embryos used for analysis therefore should be selected and the abnormal individuals discarded.

RESPIRATION OF THE FROG EGG DURING DEVELOPMENT

Since the glycogen, lipid, and protein of the developing egg are present in such high concentrations that it becomes difficult to detect any utilization of these substances during early development, we will begin our analysis with respiration. A relatively large number of measurements of the rate of oxygen consumption at various stages of development have been made. These determinations are in agreement with regard to the general shape of the curve relating the rate of oxygen consumption to the time of development. The rate of respiration increases with development as shown in Figure 2. There is also general agreement that the curve is not a smooth curve but contains plateaus over short intervals. The number of these plateaus and the time intervals over which they occur are not agreed upon at present.

The rise in respiration that occurs with growth and development of the egg is directly related to the processes of growth and development. Figure 2 presents the rate of oxygen consumption of eggs which, although they do not grow and develop beyond the early gastrula, stage 10, survive until about stage 20 when the normal embryo hatches from the egg membranes and egg jelly. The "blocked" gastrulae are obtained by fertilizing *Rana pipiens* eggs with *Rana sylvatica* sperm. After fertilization the development of these hybrids appears normal until about stage

10; they then undergo some cellular rearrangements but show no signs of differentiation (Moore, 1946). As can be seen from the curve marked "hybrid" in Figure 2, the increasing rate of respiration is blocked at or before stage 10 simultaneously with

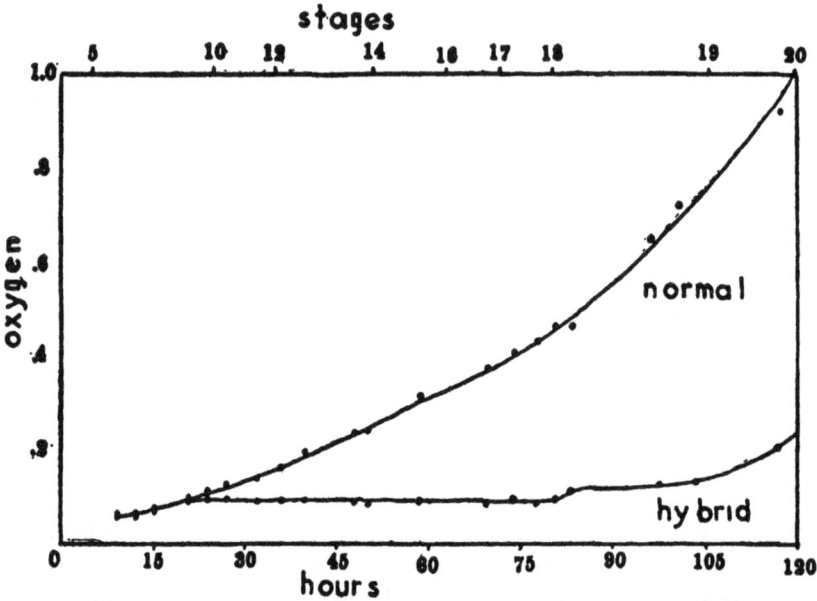

FIGURE 2. THE INCREASE IN THE RATE OF OXYGEN CONSUMPTION WITH INCREASE IN TIME OF DEVELOPMENT OF THE FROG EGG
Two curves are shown. The normal curve describes the rate of respiration of *Rana pipiens*; the hybrid curve, that of *Rana pipiens* eggs fertilized by *Rana sylvatica* sperm. These hybrid eggs develop until the beginning of gastrulation, stage 10, and then fail to show further differentiation. The temperature of the measurements was 18.7° C., and the Warburg type of manometer was used to measure the amount of oxygen consumed. A single batch of eggs was used for all the measurements at different stages of development. The oxygen is plotted in terms of mm.^{3} O_2 per hour per egg.

the developmental block. Between stages 10 and 20 of the normal embryos little or no growth and development occur in the hybrids and there is little or no increase in rate of respiration. Let us assume that the blocked hybrids represent a maintenance metabolism. The difference between the normal and the hybrid rates of respiration then is a measure of the metabolism of growth and development. How much of the latter metabolism is related to growth and how much to development?

The growth of the egg possibly might be measured by the increase in number of cells. The increase in cell number however is of a different order of magnitude than the increase in respiration. Sze (1953a) found that the gastrula at stage 10 contains 32,000 cells, while at stage 19 the embryo contains 440,000 cells. This change amounts to a 14-fold increase. Over the same interval respiration changes from about 0.1 to about 0.7, or a sevenfold increase. There is definitely a greater increase in the number of cells than in the rate of respiration. As a matter of general observation the egg cytoplasm cannot possibly increase by the same numerical value as does the cell number, for that would mean a 440,000-fold increase from fertilization to stage 19, whereas the amount of cytoplasm in the fertilized egg is estimated at about 16 percent by volume. If this estimate is anywhere near correct only a six- to sevenfold increase in cytoplasm could occur during development. The plain fact of the matter is that the nucleoplasmic ratio increases and cells become smaller with cell division.

Since we have no good estimate of the growth of the cytoplasm perhaps the growth of the nuclei as measured by desoxyribosenucleic acid (d.n.a.) might give us an estimate of growth. Referring again to Sze (1953a) we find that between stage 10 and stage 19 the d.n.a. increases from 1.3 to 6.1 micrograms per egg. This is almost a fivefold increase and may be compared with a sevenfold increase in the rate of respiration. Since active differentiation is occurring during part of the period between stage 10 and stage 19, it may be that the discrepancy between the d.n.a. measure of the growth and the actual respiratory rates is occasioned by the added respiration necessary for differentiation. Thus assuming a value of respiration of 0.1 for maintenance and increasing this by 5×, making a total of 0.6, there is left about 0.1 as a value expressing the rate of respiration which went into the process of differentiation.

Sze's figures may correlate also with the figures for respiration in another respect. It has been shown by a number of investigators (Brachet, 1950) that during neurulation there is a break in the curve of the rate of respiration with time. The stages used by Sze are rather widely spaced for drawing a curve, but if one

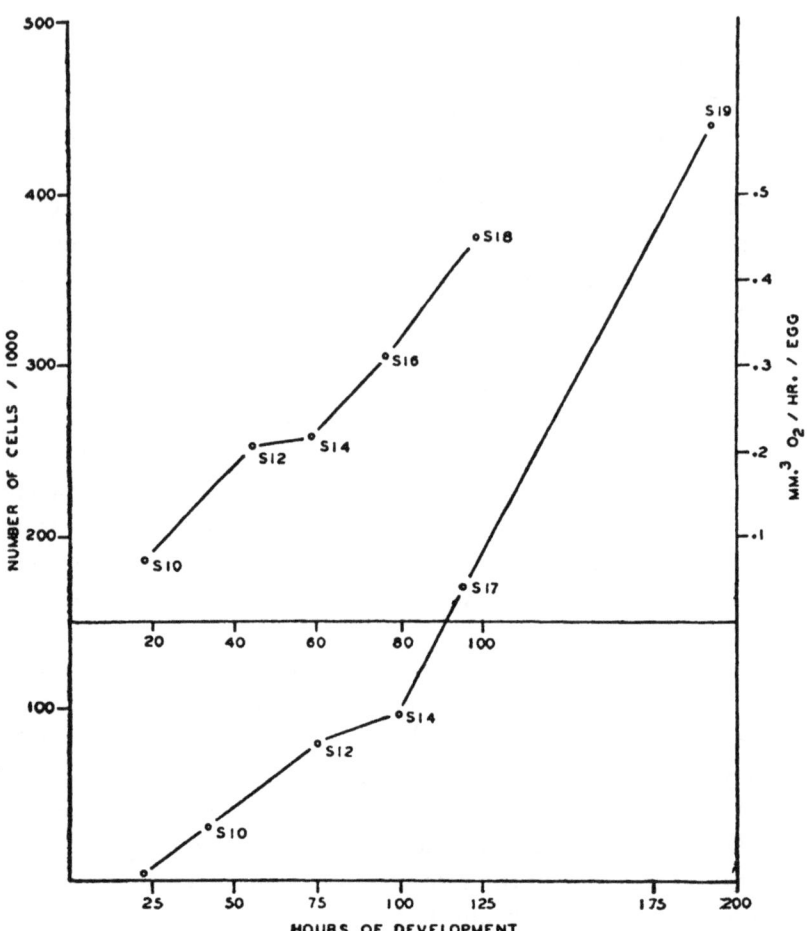

FIGURE 3. THE RELATION OF CELL NUMBER TO RATE OF RESPIRATION DURING DEVELOPMENT

A graph consisting of the number of cells in the frog embryo plotted against time is compared with a graph of the rate of respiration plotted against time. Since the two experiments were carried out at different temperatures the time axes for morphological stages do not coincide. Values for cell numbers from Sze (1953a).

plots total numbers of cells against time he obtains a curve which shows a rapid increase in cell number up to stage 12, a distinctly less rapid increase from stage 12 to stage 14, and a more rapid increase in cell number between stages 14 and 19. Thus during the period when the rate of respiration increases only slightly, the number of cells also increases only slightly.

The relationship between the rate of respiration and the number of cells in the embryo to time for development is given in Figure 3. Here Sze's data on cell number are plotted and compared with some unpublished data of our own on rate of respiration. Since the two experiments were carried out at different temperatures the morphological stages are given for both curves.

Cleavage and blastulation of the frog egg are accompanied by a steadily increasing rate of respiration. As yet there is no conclusive evidence as to the nature of the compounds oxidized. Direct chemical analyses of the lipid, carbohydrate, and protein show no significant changes between fertilization and the beginning of gastrulation. We must turn therefore to indirect methods. Measurements of oxygen consumption and carbon dioxide production give respiratory quotients which sometimes provide a clue to the nature of the substrate oxidized. The R.Q. for the period of stages 7 to 9 is 0.87, a value consistent with several possibilities in regard to substrates. More knowledge of the excretory products of the egg must be gained before any conclusions can be drawn.

CARBOHYDRATE METABOLISM OF THE FROG EGG
DURING DEVELOPMENT

The oxygen consumed during early development may be related in a large part to the carbohydrate oxidized. Thus Gregg (1948) found that the total carbohydrate changed from an initial value at fertilization of 104 micrograms per egg to 58 micrograms per hatching embryo at stage 20. These values are means of a number of experimental determinations. The difference between these two stages amounts to 46 micrograms of carbohydrate. Most of this loss is a result of a decrease in the amount of glycogen, which is present in the fertilized egg in the amount of 73.2 micrograms. The loss in glycogen is 42.2 micrograms, which may be compared with the loss of 46 micrograms of carbohydrate. Glycogen therefore is the first energy-rich compound to be utilized by the developing egg. That is to say, it is the first compound which by chemical methods thus far has been found to be converted to some nonreducing compound.

The usual course of breakdown of glycogen involves a long

series of reactions including phosphorylation, fragmentation, mutation, and oxidations coupled with phosphorylation, resulting in a three-carbon acid, pyruvic acid. The pyruvic acid then may be further oxidized in stepwise fashion to carbon dioxide and water. Under anaerobic conditions lactic acid may be formed from pyruvic acid by a reduction. Finally the pyruvic acid may participate in a wide variety of reactions and be incorporated into proteins or fats.

TABLE 5

RESPIRATION AND ANAEROBIC LACTIC ACID PRODUCTION IN THE FROG EGG

The time of development is given in terms of minutes after first cleavage, M.F.C., and the morphological stage, from Shumway, 1940. Oxygen consumption and lactic acid production are expressed as microliters gas per milligram dry weight per hour. From Cohen, 1953.

M.F.C.	Stage	Oxygen consumption	Lactic acid production in oxygen	Lactic acid production in nitrogen	Meyerhof quotient
200	6+	0.054	0.010	0.201	3.5
600	8+	0.069	0.010	0.247	3.5
1000	9	0.108	0.010	0.287	2.6
1400	11−	0.141	0.010	0.327	2.3
1900	12+	0.162	0.010	0.379	2.3

Some information on the nature of the glycogen metabolism of the frog egg is gleaned from studies involving anaerobiosis. Cohen (1953) has investigated the lactic acid production under anaerobic conditions and has found that in all the early stages of development, cleavage through hatching, lactic acid forms. Lactic acid production is inhibited almost completely by the presence of oxygen and only traces are found under normal aerobic conditions (Table 5). Thus the usual Pasteur effect of muscle metabolism is found also in the frog egg. Further, the egg will develop under anaerobic conditions, just as muscular contraction will occur in the absence of oxygen, showing that the energy of development like that for muscular contraction may be derived from the reactions occurring during anaerobic glycolysis.

In Table 5 the rates of oxygen consumption and of lactic acid production under anaerobic conditions are compared for the same clutch of eggs. The rate of oxygen consumption increases from a value of 0.054 during early cleavage to one of 0.162 at the end of gastrulation. This is a threefold increase. The rate of production of lactic acid under anaerobic conditions changes from 0.2 to about 0.4, or a twofold increase for the same interval during which oxygen consumption increased by a factor of 3. Thus the rate of oxygen consumption increases more rapidly with development of the egg than does the rate of lactic acid production.

Probably not all the oxygen consumption is a result of the oxidation of carbohydrates and thus there would not need to be a close correlation between rates of oxygen consumption and lactic acid production. Then too there may be some limiting factor in the determination of the rate of lactic acid production. Without further analysis it is not possible to explain the lack of proportionality between the rates of oxygen consumption and of lactic acid production.

TABLE 6

THE REOXIDATION OF LACTIC ACID IN THE FROG GASTRULA

After exposure to nitrogen for a period of time the lactic acid was measured and the gastrulae were exposed to air. For each determination 100 stage-10 gastrulae were used. Lactic acid was determined by oxidation to carbon dioxide.

Treatment	LACTIC ACID IN MM.^{3} CO_2			
	Exp. 1	Exp. 2	Exp. 3	Exp. 4
2 hr. N_2	51			
4 hr. N_2	94	65	62	
4 hr. N_2 + 4 hr. air	84	48	50	
8 hr. N_2	182	100	127	
9 hr. N_2				170
9 hr. N_2 + 13 hr. air				73

THE REOXIDATION OF LACTIC ACID   The lactic acid produced by the gastrula under anaerobic conditions does not leave the gastrula and is reoxidized slowly in air (Table 6). At the end of 4 hours in nitrogen (experiment 1) a value of 94 cubic millimeters of carbon dioxide was obtained for the lactic acid present in the gastrulae. When these gastrulae were returned to air for 4 hours the lactic acid dropped to 84. Gastrulae which were exposed to nitrogen for the entire 8-hour period contained 182 units of lactic acid expressed as carbon dioxide. Experiments 2 and 3 record similar changes in lactic acid. Experiment 4 shows that an appreciable amount of the lactic acid produced in a 9-hour anaerobic period is reoxidized during a 13-hour period in air. The respiratory quotient of the gastrulae during the period of reoxidation of the lactic acid was 0.95. The rate of oxygen consumption of the gastrulae during lactate oxidation is somewhat lower than that of control gastrulae kept in air continuously. The control gastrulae at the end of the experiment were slightly more advanced in development than those treated with nitrogen (stage 12+ compared with stage 12). In this type of experiment there was no evidence of an extra consumption of oxygen in gastrulae first exposed to nitrogen and then to air. During the first hour of measurement the rate of oxygen consumption was 12.7 cubic millimeters per hour per 100 eggs, while that of the control was 12.6.

The oxygen debt incurred by the egg during a sojourn at low oxygen tension is paid off slowly (Figure 4). Preliminary experiments showed that in about 5 percent air : 95 percent nitrogen the egg would develop, but the normal increase in the rate of oxygen consumption did not take place. Eggs placed in such an atmosphere at stage 8− (Figure 4) develop at about the same rate as controls and after 19 hours both are in stage 11+. The controls in air respire at a rate of 0.16 mm.^{3} of oxygen per egg per hour, while the eggs in 5 percent air respire at about 0.05. Thus a considerable oxygen debt, represented by the area between the two curves, is accumulated. The 5 percent air is replaced with 100 percent air at about 19 hours (indicated by arrow) and the rate of oxygen consumption is determined. The rate climbs steadily but does not reach the rate of controls in air for about 19 hours (38 hours in Figure 4). In four such experiments no sudden burst of oxygen consumption was obtained. Up to about 30 to 32 hours of development no easily detectable differences in the stage of development of experimentals and controls was found. At 44 hours the experimental eggs were in stage 14+, while the controls were in stage 15. The experimental neurulae were respiring at a greater rate than the controls. Thus after about 38 hours the oxygen debt began to be paid.

Similar results were obtained when nitrogen was substituted for 5 percent air, but in this case there was distinct inhibition of development during 19 hours of anaerobiosis and oxygen consumption was negligible. Upon return to air the respiration of the eggs rose sharply but did not reach that of the controls. By the time the treated eggs reached stage 13 they were respiring at a greater rate than control eggs at stage 13.

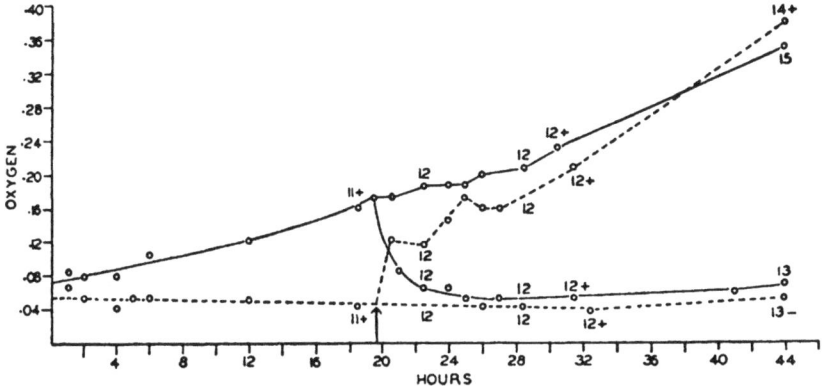

FIGURE 4. THE OXYGEN CONSUMPTION OF FROGS' EGGS IN AIR AFTER EXPOSURE TO LOW OXYGEN TENSION

All eggs at the beginning of the experiment are in stage 8—. Two groups of stage 8— eggs are in flasks containing 100% air and 2 groups are exposed to 5% air : 95% nitrogen. Solid line shows rate of oxygen consumption per egg per hour in 100% air. Broken line records rate of oxygen consumption of eggs in 5% air. At 19 hours one group of the eggs that had been in 100% air was transferred to 5% air. Similarly, 1 group of the eggs in 5% air was treated with 100% air. From 19 to 44 hours the uppermost solid line records rates of oxygen consumption of eggs continuously exposed to 100% air. Below this curve a broken line describes the rate of oxygen consumption of eggs previously exposed to 5% air but now in 100% air. Next curve below is a solid line of the rate of oxygen consumption of eggs previously exposed to 100% air but now in 5% air. Finally, the lowermost broken line represents rates of oxygen consumption of eggs continuously exposed to 5% air. The numbers near the curves are the stages of development.

Figure 4 shows the effect of reduced oxygen tension on eggs which have been in air up to stage 11+. When the air is replaced by 5 percent air : 95 percent nitrogen, respiration falls to a level slightly above that of eggs which were exposed to low oxygen tension at stage 8—. At 44 hours the eggs continuously respiring in a concentration of 5 percent air are in stage 13—, while those exposed to low oxygen tension from stage 11+ on are in stage 13.

When each of these groups of eggs is allowed to respire in 100 percent air the oxygen consumption rises and approaches that of control eggs at the same stage of development. Again there is no evidence of an increased oxygen consumption after exposure to low oxygen tensions, even after 44 hours' exposure time. The experiments unfortunately were not carried far enough to see if eventually the oxygen debt was repaid. The experiments do show that eggs may develop at a normal rate from stage 8− to 12+ although they consume much less oxygen than normally. After eggs are returned to air, development still occurs with less oxygen consumed since some of the observed oxygen consumed must be used to reoxidize the lactic acid which accumulated at low oxygen tensions.

COUPLING BETWEEN PHOSPHORYLATION, RESPIRATION, AND CARBOHYDRATE BREAKDOWN   It seems significant to call attention to the fact that lactic acid is produced under anaerobic conditions during early cleavage when no detectable carbohydrate, lipid, or protein losses can be determined by analysis. Since lactic acid is produced, however, carbohydrate probably is oxidized during this early period.

The facts that glycogen is utilized in early development (at least at stage 12) and that lactic acid is produced under anaerobic conditions suggest the hypothesis that a glycolysis similar to that of muscle occurs in the frog egg. If so, do phosphorylation and dephosphorylation couple with the carbohydrate metabolism? The presence of acid-soluble labile organic phosphates has been shown for the unfertilized egg. During early development there is little change in the concentration of this fraction. When the egg is subjected to anaerobic conditions, however, a decrease in the labile organic acid-soluble fraction is observed (Barth and Jaeger, 1947). The egg continues to develop for several hours under anaerobic conditions with the production of lactic acid, as we have seen. If the energy for development comes in part from carbohydrate metabolism through the mediation of energy-rich phosphates as in muscle, then less energy would be available under anaerobic conditions and one might expect a decrease in energy-rich phosphates.

The expected decrease of labile phosphate under anaerobic conditions is found in Table 7 where the inorganic phosphate and labile phosphate fractions of frogs' eggs kept in air and in nitrogen are compared. In nitrogen 8.7 micrograms of labile phosphate per 100 eggs are lost and 10.9 micrograms of inorganic phosphate appear. Although no identification of the labile phosphate fraction has been made, we assume that it is comprised mainly of adenosine triphosphate.

TABLE 7

THE DECREASE IN LABILE PHOSPHATE DURING ANAEROBIOSIS OF THE FROG EGG

The values are in micrograms per 100 eggs. Labile phosphate is that phosphorus which appears as inorganic phosphate after a 7-minute hydrolysis at 100° C. in 1.0 N HCl. The compound hydrolyzed is thought to be adenosine triphosphate. The eggs were early gastrulae (stage 10) and were exposed to air and nitrogen for 22 hours at 18° C. After this time the eggs in air were in stage 12, while those in nitrogen were in stage 11. From Barth and Jaeger, 1947.

Eggs in	Total acid-soluble phosphate	Inorganic phosphate	Labile phosphate
Air	49.1	24.5	24.6
Nitrogen	51.3	35.4	15.9
Difference	+2.2	+10.9	−8.7

We interpret the decrease of labile phosphates as a result of the decrease in the release of energy under anaerobic conditions. Under anaerobiosis the demands for energy are the same as for eggs developing in air, namely energy for maintenance, growth, and development. Since lactic acid forms when the eggs are developing in nitrogen, the energy of the oxidation of lactic acid to carbon dioxide and water is lacking. Thus some other source of energy must substitute for this deficit in energy. We assume that the energy comes from the high-energy labile phosphates present in the egg at the beginning of the sojourn in nitrogen.

The presence of air with the resultant complete oxidation of the carbohydrates to carbon dioxide and water allows for the synthesis of a higher concentration of labile phosphates. The actual synthesis in air may be demonstrated by first permitting most of the labile phosphate to break down in the presence of nitrogen and then admitting air to the eggs once more. As Table

8 shows, the labile phosphate of eggs in nitrogen decreases from 16.7 to 4.1 micrograms per 100 eggs. Upon admitting air to the anaerobic eggs the labile phosphate rises from 4.1 to 15.9 micrograms per 100 eggs in 2 hours. After 8 hours treatment with air the originally anaerobic eggs contain 17.3 micrograms of labile phosphate as compared with 16.7 micrograms in eggs that have been in air continuously.

TABLE 8

DEPHOSPHORYLATION AND REPHOSPHORYLATION OF
LABILE PHOSPHATES IN THE FROG EGG

Eggs are exposed to nitrogen for 22 hours and then to air for 2, 4, and 8 hours. Temperature, 18° C. The numbers signify micrograms per 100 eggs. The eggs were early gastrulae (stage 10) at the beginning of the experiment. From Barth and Jaeger, 1947.

Eggs in	Total acid-soluble P	Inorganic P	Labile P
Air, 22 hr.	45.9	29.2	16.7
Nitrogen, 22 hr.	41.9	37.8	4.1
Nitrogen, 22 hr. + air, 2 hr.	48.5	32.6	15.9
Nitrogen, 22 hr. + air, 4 hr.	48.3	27.7	20.6
Nitrogen, 22 hr. + air, 8 hr.	43.2	25.9	17.3

The labile phosphate lost during anaerobiosis thus is regained in air. The increased energy production in the presence of air makes possible a synthesis of the energy-rich phosphate from inorganic phosphate and adenosine diphosphate. For, while the breakdown of labile phosphates might be completely independent of carbohydrate metabolism since these labile phosphates lose phosphate with a decrease in free energy, the synthesis of labile phosphate from inorganic phosphate as one of the reactants can occur only by means of a coupled energy source. The possibility of an energy source of nonrespiratory origin with high enough potential to couple with the formation of a labile phosphate bond is unlikely. We conclude therefore that carbohydrate breakdown in the frog egg is coupled with phosphorylation. This does not mean however that phosphorylation and dephosphorylation are further coupled with development, as the energy of the phosphate bond could be concerned merely with maintenance.

We thought that a study of the hybrid gastrulae which do

not develop but nevertheless are maintained as gastrulae might provide some further information. A comparison of the effects of anaerobiosis upon normal and hybrid gastrulae was made. Briefly, the comparison showed that labile phosphates break down and inorganic phosphate appears in the arrested hybrid gastrula as well as in the normal, developing gastrula. There is, in fact, a slightly greater loss of labile phosphate in the hybrid gastrula under anaerobic conditions. Thus it could not be demonstrated that the process of development was correlated with the breakdown of labile phosphate. With no development occurring in the hybrid eggs labile phosphate broke down.

In sum, then, we have made reasonable progress toward revealing a coupling between phosphorylation and respiration, but the coupling between dephosphorylation and development is still obscure.

GLYCOLYSIS IN HOMOGENATES OF THE EGG   To continue with the general nature of carbohydrate metabolism of the frog egg we turn to some studies on the lactic acid production in homogenates of eggs (Cohen, 1953). Although homogenates respire readily (Spiegelman and Steinbach, 1945) and even exceed the normal rate of respiration of the intact egg, no lactic acid is produced by egg homogenates under anaerobic conditions. Some members of the chain of reactions leading to the appearance of lactic acid probably are inactivated by homogenizing. Cohen succeeded in obtaining lactic acid from egg homogenates by adding magnesium chloride, the potassium salt of adenosine triphosphate, diphosphopyridine nucleotide, and various hexose phosphates as substrates. Glucose would not serve as a substrate, but glycogen with added inorganic phosphate was converted to lactic acid. Figure 5 illustrates the lactic acid production in a homogenate containing a constant amount of glycogen but in which the concentration of inorganic phosphate is varied. We see that, while a small amount of lactic acid is produced in a brei containing no added phosphate, the amount of lactic acid formed is much greater when inorganic phosphate is added. With conditions optimal for lactic acid formation none is formed if glycogen is absent.

It is clear then that the egg contains all the enzymes necessary for the conversion of glycogen into lactic acid. Since glucose-1-phosphate and hexose-di-phosphate each will serve as a substrate for lactic acid production, all the enzymes are present for the conversion of glycogen into hexose-di-phosphate. With

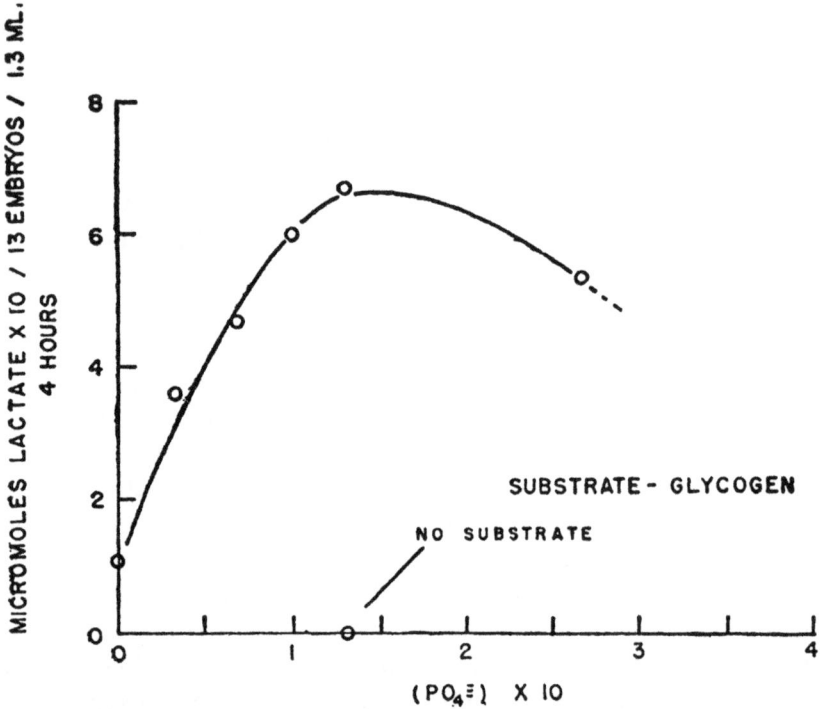

FIGURE 5. THE FORMATION OF LACTIC ACID FROM GLYCOGEN BY THE ACTION OF THE ENZYMES IN HOMOGENATES OF FROG GASTRULAE
The amount of acid produced in 4 hours depends upon the concentration of inorganic phosphate. At the optimal concentration of inorganic phosphate for lactic acid formation none appears unless glycogen is present. From Cohen, 1953.

this information in mind we may recall now that the cleaving egg produces lactic acid and tentatively conclude that the lactic acid was derived from glycogen. Actually the homogenate studies show that the egg can produce lactic acid from glycogen. They do not prove that the living egg does so.

One further step in the argument for the origin of lactic acid from glycogen in cleavage stages comes from Cohen's studies on the different stages of development (Table 9). In these experiments the eggs were homogenized at stages from first cleavage to the end of gastrulation and the amount of lactic acid formed from glycogen was determined for a 2-hour period in each case. We see that the enzymes of the egg at first cleavage are able to convert glycogen into lactic acid. Thus even before any measurable decrease of glycogen occurs in the egg the enzymes are present for its conversion to lactic acid. Since the intact egg in early cleavage produces lactic acid it seems highly probable that this lactic acid comes from glycogen.

TABLE 9

LACTIC ACID PRODUCTION FROM GLYCOGEN IN HOMOGENATES OF EGGS AT VARIOUS STAGES OF DEVELOPMENT

Lactic acid is measured as the total amount in micrograms produced by a homogenate of 15 individuals in two hours. Temperature of development, 20° C. From Cohen, 1953.

Eggs in	Stage	Minutes after First Cleavage	Lactic Acid
First Cleavage	3	20	21
Cleavage	7	360	14
Cleavage	8+	720	23
Early Gastrula	10	1200	29
Late Gastrula	11+	1600	27
End Gastrulation	12+	2000	22

We are again confronted, however, with a difficulty of interpretation. The homogenate studies show that an egg can convert glycogen into lactic acid but they do not prove that the conversion actually occurs in the living egg. Tables 5 and 9 illustrate this point. The amount of lactic acid produced at first cleavage is equal to that produced at the end of gastrulation. Table 5 however shows that the amount of lactic acid produced in the living egg is about twice as much at gastrulation as compared with cleavage. The oxygen consumption is three times as high at stage 12+ as in cleaving eggs. Yet in homogenates the differences between cleaving eggs and gastrulae no longer appear (Table 9). In the living egg the structural relationships between enzymes, coenzymes, and glycogen must be such that not all the enzymes or coenzymes are able to unite with glycogen. As development proceeds more and more glycogen comes in contact with the enzyme-coenzyme system and a higher rate of lactic acid production results.

Cohen's studies demonstrate that lactic acid is produced in homogenates from glycogen, glucose-1-phosphate, glucose-6-phosphate, fructose-6-phosphate, and hexose-di-phosphate. What are the intervening steps between hexose-di-phosphate and the appearance of lactic acid? Brachet (1950) reports with Rapkine the presence of triosephosphate dehydrogenase in the frog egg, while Lindahl and Lennerstrand (1942) measured the activity of coenzyme in various stages of development. Thus two additional compounds are present and this suggests that the hexose-di-phosphate is converted to a triosephosphate, which in turn is oxidized to phosphoglyceric acid. Further speculation awaits a more complete study of the intermediary metabolism of the carbohydrates in the frog egg.

THE EFFECT OF INHIBITORS ON DEVELOPMENT AND METABOLISM

Some additional information may be obtained from studies of the effects of various compounds which are known to inhibit or stimulate various steps in metabolism. Sodium fluoride, for example, inhibits the action of enolase which catalyzes the reaction: 2-phosphoglyceric acid into phospho-enol-pyruvic acid. If then sodium fluoride applied to the frog egg inhibits respiration and glycolysis its effect is evidence, but not proof, that enolase is present in the egg.

At least three major difficulties are encountered in a study of the effect of inhibitors on development and metabolism of the frog egg. The frog egg is relatively impermeable and substances, even water, enter and leave the egg slowly. Thus if a metabolic inhibitor has no observable effect it cannot be concluded that the enzyme which the inhibitor poisons is lacking. The alternative interpretation that the inhibitor fails to penetrate is the more likely conclusion.

Another difficulty stems from the fact that metabolism of the egg increases with development. If the respiration of eggs in an inhibitor is lower than the respiration of controls is this difference a result of a difference in rate of development, or is the difference in respirations due to an inhibition of metabolism by the inhibitor? In other words, does the inhibitor act upon respiration and therefore retard development or does it inhibit development and therefore prevent the normal increase in metabolism?

Finally, a general criticism of the use of inhibitors to detect specific enzymes has been made. Rarely in a system of mixed enzymes such as that found in the living egg does a respiratory inhibitor inhibit one and only one enzyme. It is more likely to inhibit several different enzymes to different degrees. The exact interpretation of the inhibition of respiration or of lactic acid production by any substance is not simple. Taken along with other studies, however, the investigations of the effect of inhibitors on metabolism and development are useful.

In some studies presented in abstract about 1947, it was found that sodium fluoride had little or no effect upon respiration and glycolysis of the frog egg. In 0.025 M sodium fluoride, a concentration that inhibits the catalysis by enolase, eggs at stage 9 respired for 19 hours at the same rate as control eggs. By this time the normal eggs were in stage 12 and the eggs in sodium fluoride were in about the same stage as the controls. Later cytolysis began in the sodium fluoride eggs, but respiration was not measured under these conditions. Experiments in which stage 12 was treated with sodium fluoride gave similar results for 8 hours of respiration. After this period respiration and development both were slowed in sodium fluoride and the treated eggs began to show cytolysis at stage 14, at which time the controls were in stage 15.

The effect of sodium fluoride upon anaerobic glycolysis was determined by two methods. In the first series of experiments the liberation of carbon dioxide was measured under anaerobic conditions as an indicator of the formation of lactic acid in the egg. The frog egg contains bound carbon dioxide from which carbon dioxide is liberated by an acid. The amount of bound carbon dioxide increases during aerobic development to stage 13, after which the amount decreases. Under anaerobic conditions lactic acid forms within the egg and the bound carbon dioxide is released and may be measured manometrically.

A typical experiment illustrating the relationships between lactic acid and the bound carbon dioxide of the egg is shown in Figure 6. Eggs in stage 10 were subjected to nitrogen and the carbon dioxide liberation was measured at various times. At the same time the eggs were analyzed for lactic acid. Since completely anaerobic conditions were not attained, oxygen consumption and carbon dioxide production also were measured.

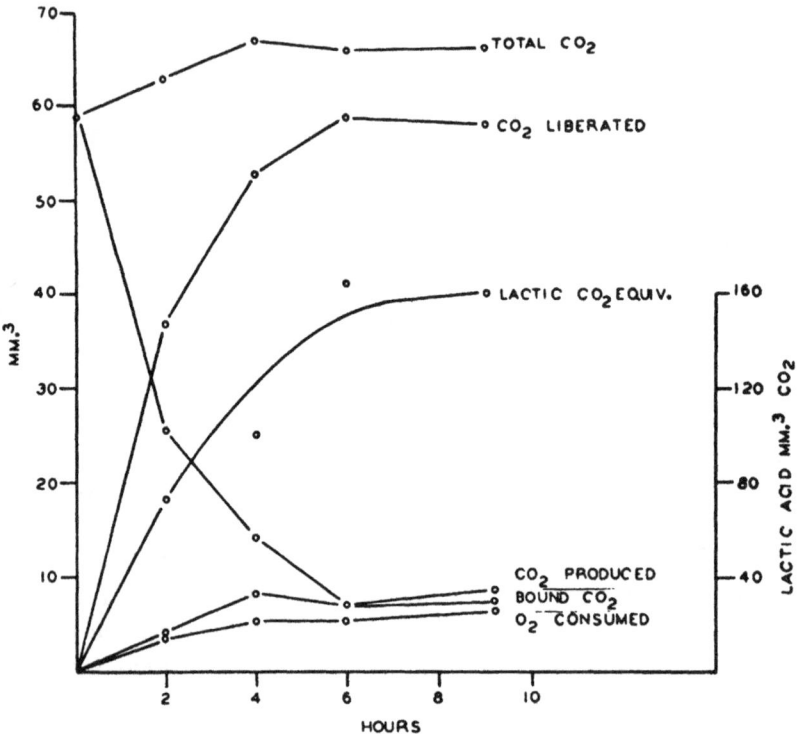

FIGURE 6. THE RELATIONSHIP BETWEEN LACTIC ACID AND BOUND CARBON DIOXIDE OF THE EGG DURING ANAEROBIOSIS
Eggs in stage 10 were subjected to nitrogen and the carbon dioxide liberation and lactic acid content were measured at various times. Oxygen consumption and carbon dioxide production also were measured.

As lactic acid is formed in the egg the bound carbon dioxide is liberated until at about 9 hours the lactic acid production falls off and no more bound carbon dioxide is liberated.

A number of experiments in which the liberation of carbon dioxide from normal eggs was compared with the liberation of carbon dioxide from sodium fluoride treated eggs revealed no differences. Stage 2, stage 8, stage 9, stage 10, and stage 12 were tested, and in each case the sodium fluoride did not alter the rate of liberation of carbon dioxide. In all cases measurements were made before the treated eggs were retarded and before cytolysis set in.

Since the liberation of carbon dioxide is only an indicator of lactic acid production the foregoing experiments were checked by a direct analysis of lactic acid. In one experiment the lactic acid formed by stage 10 eggs during 7 hours of anaerobiosis in 0.025 M sodium fluoride was 0.65 mg. per 100 eggs, while in the controls the value was 0.68 mg. In another experiment the eggs were treated with sodium fluoride for 17 hours before being subjected to anaerobiosis in order to give the fluoride a chance to penetrate. During a 9-hour period of anaerobiosis stage 10 eggs in fluoride produced 0.67 mg. of lactic acid, while the controls formed 0.72 mg. In both experiments half-hour readings of the carbon dioxide liberated failed to show any differences between the manometers containing sodium-fluoride-treated eggs as compared with manometers containing control eggs. The direct chemical analysis of lactic acid in sodium-fluoride-treated eggs and in normal eggs thus bear out the findings from a study of carbon dioxide liberation. There is no effect on anaerobic glycolysis, at least prior to the time when morphological effects become noticeable.

The above findings agree with Brachet (1950) who reports that the respiration of the egg of *Rana fusca* is not affected by sodium fluoride. Recently Ornstein and Gregg (1952) measured the amount of oxygen consumed by dorsal and ventral explants of the gastrula, stage 10, of *Rana pipiens*. They found that 0.02 M sodium fluoride reduces respiration of stage 10 explants by about 40 percent. This is a highly significant reduction in oxygen consumption and would have been detected easily by previous investigators who used the entire egg. The difference between the action of sodium fluoride on the entire egg and on an explant may be a result of the impermeability of the whole egg and the relative permeability of an explant. The whole egg is covered with a surface coat, while the explant has an inner (blastocoel) surface which lacks the surface coat. There are indications that substances do not penetrate easily through the surface coat layer but will penetrate through the blastocoel surface. For example, lactic acid produced by the egg does not leave the egg and is not found in the medium around the egg. Lactic acid does leave the cells and pass through the inner (blastocoel) surface and get into the blastocoel. The presence of lactic acid in the blastocoel is evidenced by the fact that the bound carbon dioxide of the blastocoel fluid (Gregg and Ballentine, 1946) is liberated during lactic acid production.

Sodium monoiodoacetate in equilibrium with monoiodoacetic acid acts very much as does sodium fluoride except that the action of the former compounds depends upon the hydrogen ion concentration of the mixture. A preliminary study of the morphological effects of sodium monoiodoacetate at various hydrogen ion concentrations showed that while 0.009 M iodoacetate at pH 8.6 showed no visible action for 24 hours, at pH 5.1 it caused cytolysis in 6 hours in eggs at stage 12. At pH 5.4 the 0.009 M iodoacetate caused a distinct retardation of development by 18 hours and cytolysis at 24 hours. The effects of monoiodoacetate are not caused by the acetate ion, as controls with acetic acid give normal development for at least 48 hours, after which the embryos were not examined.

With regard to anaerobic glycolysis and oxygen consumption no differences between iodoacetate-treated eggs and control eggs could be detected for the first 8 hours for stages 9, 10, and 12. The concentration was 0.009 M at pH 5.4, and cytolysis occurred in 24 hours. Thus either the iodoacetate does not penetrate the egg in 8 hours or the egg contains no respiratory enzymes which are inhibited by 0.009 M iodoacetate.

Evidence that the iodoacetate probably does not penetrate rapidly at pH 5.4 comes from experiments in which the pH was lowered to 5.1 but the concentration kept at 0.009 M. Stage 13 eggs placed in this solution were compared with controls for oxygen consumption and anaerobic release of carbon dioxide. The R.Q. also was determined for the eggs in monoiodoacetate. The oxygen consumption and release of carbon dioxide was the same in the treated and normal eggs for one hour. The oxygen consumption of the treated eggs then declined steadily, and at

5 hours the rate for the eggs in iodoacetate was about 40 percent of the normal rate of respiration. The amount of carbon dioxide released from the egg under anaerobic conditions after one hour in iodoacetate began to increase and greatly exceeded that of the controls. This increase in carbon dioxide liberated was possibly a result of the penetration of monoiodoacetic acid at pH 5.1. The acid reacted with the bound carbon dioxide of the egg and liberated free carbon dioxide. Thus both the change in oxygen consumption and in carbon dioxide liberation show that the acid penetrates after about one hour. With the penetration of the acid a 40 percent lowering of the rate of respiration occurs. This could be interpreted to mean that monoiodoacetate reacts with triosephosphate dehydrogenase and inhibits this enzyme. The failure to detect a lowering of respiration in other instances may be assigned to the lack of penetration of the iodoacetate. Finally, after 5 hours' treatment the eggs in iodoacetate begin to exhibit cytolysis. The decrease in respiration thus occurred before any extensive cytolysis began and was caused probably by the direct action of iodoacetate upon the respiratory enzymes. Indeed, since cytolysis usually results in an increase in respiratory rate, the inhibitory action of monoiodoacetate could not be by means of cytolysis. The action of sodium fluoride and monoiodoacetate acid on the frog egg resulting in an inhibition of respiration supports the hypothesis of a glycolysis similar to that of muscle.

Brachet (1950) has reviewed the earlier literature on metabolic inhibitors and we have nothing more to contribute. The chief studies on the effect of inhibitors on respiration and development of *Rana pipiens* were made by Barnes (1944), Spiegelman and Moog (1945), and Pomerat and Haringa (1939).

Cohen's studies on anaerobic glycolysis of egg homogenates offer an opportunity to study in some detail the effects of various inhibitors of respiratory metabolism. The factor of permeability is eliminated and the possible indirect effect of an inhibitor upon metabolism after first inhibiting development is also out of the question. There remains only the question of the specificity of the action of the inhibitor in mixed enzyme systems. Possibly a partial fractionation of the egg homogenate into simpler systems of enzymes will help determine the locus of the action of an inhibitor.

If it becomes necessary to work with the living cells rather than homogenates in a study of inhibitors, then fragments of the egg appear to be better suited than intact eggs. The extreme impermeability of the intact eggs to salts makes it necessary to use egg fragments which, judging from Ornstein and Gregg's data on sodium fluoride, appear to be much more permeable.

THE METABOLISM OF BLOCKED HYBRID EMBRYOS   Another type of inhibitor is not subject to the uncertainties of permeability, namely the inhibitor produced by hybridization. Moore (1941, 1946, 1947) has provided the chemical embryologist with a wealth of material for investigation in this way of a variety of frog hybrids. These hybrids develop to various stages and then become arrested. A partial analysis of the respiratory metabolism of hybrid eggs has been begun, and a summary of the situation for the cross, *pipiens* female by *sylvatica* male, is to be found in a paper by Gregg (1948). The rate of oxygen consumption, the rate of lactic acid production, the respiratory quotient, the bound carbon dioxide, and the concentration of labile phosphate are the same for the hybrid egg as for the normal egg from fertilization to the beginning of gastrulation (stage 10). Carbohydrate metabolism (glycogen disappearance) may possibly begin earlier in the hybrid as compared with normal. After stage 10, at which time the hybrids are arrested, the rate of oxygen consumption and the rate of lactic acid production fail to increase as in normals. The disappearance of glycogen continues in the hybrid, but at a much slower rate than in normal eggs. Whereas the normal egg consumes about 58 percent of its glycogen between fertilization and hatching, the hybrid blocked at stage 10 consumes only 22 percent between fertilization and the time at which normal eggs hatch. Gregg concludes that there is a "partial blockage of at least one step in the glycolytic chain" of reactions. He suggests a further investigation of the reaction diphosphoglyceraldehyde plus oxidized diphosphopyridine nucleotide (D.P.N.) to give phosphoglyceric acid and reduced

D.P.N. Since this step also is concerned with the phosphorylation of adenosine diphosphate, a partial blockage at this step would account for the finding of Barth and Jaeger (1947). They reported that under anaerobic conditions the hybrids showed a decreased capacity for keeping adenosine triphosphate phosphorylated.

As far as the problem of relating respiratory metabolism with development goes, we are not in a position as yet to decide whether the block to respiratory metabolism precedes the block to development, or whether development is blocked and as a consequence no further changes in respiratory metabolism occur.

That both the block to respiration and the block to development are general throughout the egg is clear however. One might have expected a localized block at the dorsal lip region. The experiments of Moore (1947) show that the competence for neural differentiation is less in hybrid than in normal gastrulae. Sze (1953b) tested four fragments of the gastrula and found that each fragment of the hybrid respired at a lower rate than the corresponding fragment from a normal gastrula. These fragments were, roughly, the region above the dorsal lip, the presumptive neural plate, and two regions of the epidermis. Thus physiologically and morphologically speaking, the block is probably in all parts of the gastrula and certainly is not restricted to the organizer region.

Let us return for a moment to the question of the relation between developmental blocks and changes in respiration. If the block in metabolism is considered as preceding the block in development then we may envisage three possibilities. First, the block may occur in the reaction yielding energy, as Gregg has suggested. The energy supply would thus be decreased and an arrest of development might be expected. If, on the other hand, the block occurs in the energy transfer system, let us say in phosphate transfer, then the respiration may be unaffected, yet a block to development would occur as a result of a decrease in the availability of energy. Finally, the block in metabolism may occur in the terminal reactions that accept energy from the reactions which transfer energy from respiration. The respiration then will be normal, the phosphates present in full complement, but an enzyme or substrate for a reaction leading to the synthesis of a compound may be inhibited or missing. In short, metabolism may be intact, save for the final acceptor of energy.

The first of these possibilities may be realized in the *pipiens* × *sylvatica* hybrid where the energy supply is decreased definitely by the block. A case of the second or third situation may be that of the *pipiens* × *clamitans* hybrid in which development proceeds to gastrulation and then ceases. Healy (1952) found that the rate of respiration of the hybrids continues to increase in a normal fashion after the block to development. Obviously then enough energy is available for development, but it simply isn't utilized. Whether the block is at the level of energy transfer or at that of energy utilization has not yet been studied in any detail. A thorough study of the phosphorylated intermediates possibly might reveal a block in energy transfer.

A complete study of the metabolism of various frog hybrids may reveal blocks at various steps in the glycolytic phosphorylation-dephosphorylation chain of reactions.

PHOSPHATE METABOLISM DURING DEVELOPMENT

Meanwhile a noteworthy beginning has been made by Kutsky (1950) in the attack upon the problem of energy transfer during normal development. Changes in the phosphate fractions of the egg of *Rana pipiens* from stage 2 to stage 18 were measured by Kutsky. In a monumental study of the incorporation of P^{32} into the various phosphate fractions and the changes in specific activity of these fractions with development she was able to establish certain significant facts (Table 10).

The first part of the table records the various phosphate fractions as percentages of the total phosphate present. Since the total phosphate does not change during development these percentages are directly proportional to the absolute amounts determined by analysis. Thus most of the phosphate of the developing egg is present in the form of phosphoprotein. Next comes phospholipid, then nucleic acid, and finally labile phosphate and inorganic phosphate. The chief synthesis of phosphate-containing compounds is that of nucleic acids. An increase of 23 percent occurs between stage 2 and stage 18. The phosphate for nucleic acids must come from a breakdown of a phosphoprotein, since this fraction undergoes a decrease of 13 percent. Since the phosphoprotein is present in much greater amounts than the nucleic acids, this 13 percent decrease is more than enough to supply the phosphate for the 23 percent increase in nucleic acid phosphate.

That the phosphate from phosphoprotein is not donated directly to nucleic acid during synthesis of the latter is shown

TABLE 10

THE DISTRIBUTION OF PHOSPHORUS IN THE VARIOUS PHOSPHATE FRACTIONS OF THE DEVELOPING EGG OF *Rana pipiens*

Upper half: Distribution of P^{31}. Lower half: The distribution of tracer phosphorus (P^{32}) in the same fractions. Values for P^{31} are in terms of total phosphorus, which remains constant (Kutsky, 1950). Values of P^{32} are percentages of total P^{32}. From Kutsky, 1950.

	PERCENT OF TOTAL PHOSPHORUS				
Fraction	Stage 2	Stage 10	Stage 15	Stage 18	Change
P^{31}					
Phosphoprotein	61.0	63.0	58.0	53.0	−13%
Nucleic acid	13.0	12.0	15.0	16.0	+23%
Labile	2.6	2.5	2.2	2.9	—
Phospholipid	18.0	19.0	18.0	20.0	—
Inorganic	0.46	0.88	0.69	1.03	—
P^{32}					
Phosphoprotein	1.96	3.82	5.22	6.46	+230%
Nucleic acid	2.44	7.1	30.0	45.0	+1700%
Labile	39.0	37.0	16.0	14.0	−64%
Phospholipid	5.8	5.1	8.6	10.8	—
Inorganic	5.3	7.8	9.0	5.7	—
Total acid-soluble	76	81	54	33	−56%

by the studies on tracer phosphorus in the lower part of Table 10. Here the percentage of the total tracer phosphate present in the egg is recorded for each fraction. Thus it is seen that 76 percent of the tracer is in the acid-soluble fraction, and within this fraction 39 percent is present in the labile fraction. Although the tracer is inorganic phosphate, most of it ends up in the labile fraction in the egg.

On the other hand, although the phosphoprotein phosphorus makes up the bulk of the phosphate in the egg (61 percent), only about 2 percent of the tracer phosphate is present in the phosphoprotein. This indicates that the phosphoprotein is not in a very active state of metabolism. With development the amount of tracer phosphate in phosphoprotein, phospholipid, and nucleic acid increases, while that of the acid-soluble fraction decreases. This means that P^{31} from the three classes above exchanges with the P^{32} of the acid-soluble fraction. Since the P^{32} of the inorganic fraction is about the same at stage 18 as at stage 2, it is clear that the P^{32} incorporated into phosphoprotein, nucleic acid, and phospholipid must come from the labile phosphate fraction. The actual analysis by Kutsky shows a decrease of some 64 percent in the P^{32} content of the labile phosphate. This decrease is not a result of a loss in labile phosphate, since the amount present in stage 18 is about the same as that in stage 2. The inorganic tracer phosphate therefore first exchanges with P^{31} of the labile phosphates. The labile phosphates then donate phosphate to nucleic acids and finally the phosphoprotein breaks down to replenish the phosphate lost by labile phosphate.

From Table 10 we see also that with time more and more tracer phosphate exchanges with the P^{31} of both phosphoprotein and phospholipid. This steady exchange may represent a slow attainment of an equilibrium and does not indicate synthesis, since the phosphoprotein by direct analysis decreases, and the increase in phospholipid may not be a real one. The steady exchange between P^{32} and P^{31} in phosphoprotein and phospholipid could also be interpreted as an increase in availability of these compounds as a result of some physical change in the nature of the yolk platelet.

The general picture of the metabolism of the whole egg as a phosphorylating glycolysis appears to be developing from the studies presented in this chapter. The egg is composed of parts possessing different presumptive values in development, and we turn now to studies of the metabolisms of the parts of the gastrula.

four

**LOCALIZATION OF THE
RELEASE OF ENERGY**

STUDIES of the release of energy in whole eggs and in homogenates of whole eggs are very valuable for ascertaining the chemical reactions that occur when energy-rich compounds break down. They give little information, however, concerning the relation of metabolism to differentiation, for they do not tell us what goes on in the various regions of the egg. Since differentiation is a matter of localized embryological processes within the egg we must seek to determine any differences in metabolism which accompany the embryological processes.

SPECIAL DIFFICULTIES IN THE ANALYSIS OF EGG FRAGMENTS

The analysis of egg fragments presents some special difficulties. The interpretation of any measurement on a fragment is not simple, since an isolated fragment does not necessarily represent that same fragment when it is a part of the intact egg. Isolated fragments of an egg behave differently from the same fragments in the whole egg in regard to differentiation. It might be expected therefore that the metabolism of an isolated fragment would be different from the metabolism of the same region surrounded by the rest of the egg. A certain region of the dorsal lip of the blastopore, for example, normally differentiates into notochord in the intact gastrula, but an isolated fragment develops into somite, neural tube, and epidermis in addition to notochord. Is it not likely then that the metabolisms of the fragments in the two situations will be different? If one were trying to determine whether there was a difference in the metabolism between the presumptive notochord and the presumptive mesoderm he would find it disconcerting to discover that both regions formed neural tubes, notochord, somites, and epidermis!

Another difficulty in the analysis of parts of the frog gastrula arises from the presence of a relatively large amount of cell inclusions—yolk, glycogen granules, etc. If these inclusions were distributed homogeneously each fragment would have a constant percentage of them and the analysis of various fragments would measure any differences in metabolism. Unfortunately for the chemical embryologist the inclusions are dispersed in an unequal manner in various fragments of the gastrula. Any differences in metabolism which are found by analysis may therefore be artifacts. It is entirely conceivable that the cytoplasm of the frog egg is homogeneous and that all measurable differences are consequences of the presence of different amounts of cell inclusions. Indeed in the sea urchin egg where the cellular inclusions are distributed in homogeneous fashion, there are no differences in the rates of respiration of the parts of an egg (Lindahl and Holter, 1940).

The above difficulty has been recognized by Gregg and Løvtrup (1950) and the need for some means of eliminating the yolk granules as a fraction of the sample used for analysis is discussed. One method consists of an extraction of the non-yolk fraction of an explant. The total nitrogen of this fraction is determined and biochemical variables, such as activities of enzymes, are referred to this total nitrogen. The non-yolk fraction contains the cytoplasmic proteins and therefore serves as a measure of the amount of cytoplasm in any particular gastrular fragment.

There remains another variable in studies on localization of energy release within the egg. The gastrula must be dissected by hand and the parts isolated in some orderly fashion. The only sharp reference mark on the early gastrula is the dorsal lip of the blastopore. The approximate positions of the animal and vegetal poles may be judged but no physical marker exists. As a result the dissection at the best is rather crude and varies with the investigator; it is doubtful whether even a single investigator can be consistent in his dissections. If one wishes to compare the dorsal lip region with the corresponding region on the ventral side of the egg it is not easy to obtain the exactly corresponding region. If one cuts just a little too far toward the vegetal pole, he will obtain a fragment with relatively greater amount of yolk than the dorsal lip and some biochemical variables will then appear to be less in this region than in the dorsal lip. If, on the other hand, the incision is too far toward the animal pole, then relatively more protoplasm will be present as compared with the dorsal lip and some variables will measure higher than in the dorsal lip. The even greater difficulty of recovering corresponding fragments at two different times of development is obvious despite the guidance afforded by the classical gastrula maps of Vogt.

The more recent work which we will consider here has taken the difficulty of dissection seriously. The investigators have isolated a continuous series of samples from animal to vegetal pole on both the dorsal and ventral sides of the gastrula. This gives us a complete picture of differences within the egg along the animal-vegetal axis. It does not however solve the difficulty of comparing dorsal and ventral fragments. The whole series of dorsal fragments may contain too little or too much yolk with respect to the corresponding series of ventral fragments.

Another variable in dissection introduces quantitative differences in results. The size of the fragment used for analysis is important for, with a vegetal-animal gradient in the concentration of yolk, a relatively large fragment from the animal pole will contain relatively more yolk and certain enzymatic activities will be lower and certain compounds will be present in lower concentration.

The variables of location of fragments in the gastrula and of the size of the fragments can be expressed only by presenting accurate diagrams of the operation and perhaps measurements of the size of the fragments. Different investigators probably will continue to use individual methods of dissection.

RESPIRATION OF THE PARTS OF THE FROG GASTRULA

We will begin our discussion of the localization of energy release by considering the recent work on respiration of the parts of the frog gastrula by Sze (1953c). Figure 7 shows the method of dissection of the gastrula, stage 10, and the resulting fragments. In the frog gastrula the probable presumptive values (significance) of the fragments are as follows: 1, notochord; 2, neural plate; 3, epidermis; 4, epidermis and mesoderm. D and V signify dorsal and ventral halves of the yolk-laden endoderm cells and as dissected out probably contain some mesoderm. In

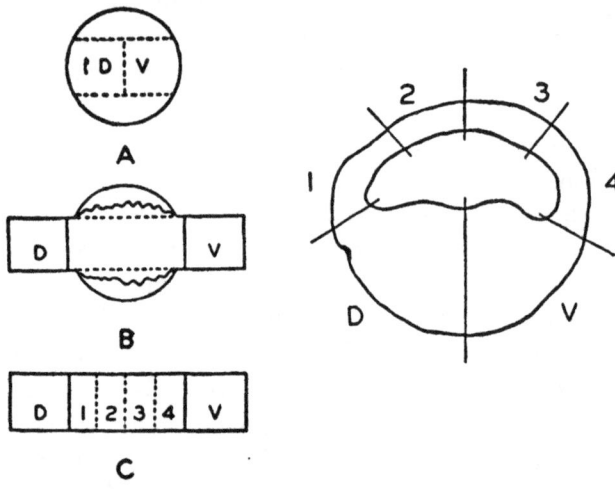

FIGURE 7. THE METHOD OF DISSECTION AND THE RESULTANT FRAGMENTS OF THE FROG GASTRULA, STAGE 10
D is dorsal; V is ventral; region 1 is just above the dorsal lip of the blastopore and is the organizer region; region 4 is the region on the ventral side corresponding to region 1 on the dorsal side. A comparison of region 1 with 4 reveals differences between the organizer region and a nonorganizer region. From Sze, 1953c.

any case, a comparison of 3, 4, and V gives information about the animal-vegetal differences. The differences between the dorsal and ventral sides of the egg are obtained by comparing 1 with 4, 2 with 3, and D with V. The rates of respiration of the gastrular fragments in terms of dry weight are recorded in Table 11. Looking at the mean values we see differences along the animal-vegetal axis of the egg by comparing regions 3, 4, and V (4.9, 3.2, 0.90, respectively). This is a result expected from our knowledge of the distribution of the yolk. Region 3 with the least yolk respires at a greater rate than region V, which contains the most yolk. The differences in rate of respiration may possibly be a result of differences in the amount of yolk.

Dorsal-ventral differences in the rate of respiration in terms of dry weight are seen by comparing fragment 1 (dorsal lip) with fragment 4 (presumptive ventral lip). The value 4.1 as compared with 3.2 shows that the dorsal, organizer region consumes more oxygen per unit of dry weight than does the nonorganizer region taken from a corresponding locus on the ventral side of the gastrula. The higher rate of respiration of the organizer region (1) as compared with nonorganizer (4), at one time

TABLE 11

THE RATES OF RESPIRATION OF FRAGMENTS OF
THE FROG GASTRULA, STAGE 10

The fragments are obtained as shown in Figure 7. The values are microliters of oxygen consumed per hour per microgram dry weight of the fragment multiplied by 10,000. From Sze, 1953c.

Egg Number	D	1	2	3	4	V
1	0.77	3.6	4.6	4.0	2.9	0.76
2	0.84	3.1	3.6	3.8	2.3	0.71
3	1.7	3.8	5.1	4.8	3.3	1.4
4	2.3	5.2	5.4	6.0	3.7	1.0
5	0.45	4.6	5.2	5.7	3.6	0.64
Mean	1.2	4.1	4.8	4.9	3.2	0.90

a matter of controversy among investigators, now appears as an established fact. Ornstein and Gregg (1952) in a precise investigation of the nature of the metabolism of the dorsal lip as compared with the presumptive ventral lip (similar to regions 1 and 4 of Sze), found the dorsal lip to respire at a higher rate than the ventral lip when the rates were calculated in terms of dry weights. The ratio of the rates, dorsal lip to ventral lip, in a series of experiments were: 1.2, 1.1, 1.1, 1.1, 1.4, 1.2, 1.4, 1.2, 1.4, 1.3. Sze (1953c) records an average ratio of 1.3. With the higher respiratory metabolism of the dorsal lip as compared with the presumptive ventral lip firmly established by Sze, we may go on to examine its significance.

A higher rate of respiration of one part as compared with another might be correlated with a higher concentration of respiring protoplasm or it might mean that the respiration of the protoplasm of one part actually is higher than in another. In one case we are simply dealing with a situation in which one part contains more protoplasm than another and the difference in respiration does not appear to have any real significance. If the protoplasm of one part actually respires at a greater rate than that of another part, then the difference may be very significant; for the extra respiration may be coupled with an extra synthesis or with some other form of work necessary for differentiation. Sze (1953c) calculated the rate of respiration of gastrular fragments per unit of dry weight, per unit of total nitrogen, per unit of dipeptidase activity, and per unit of extractable fraction. The last is a fraction which does not contain yolk and therefore is mostly cytoplasm. When the respiration per unit of time is divided by the dry weight (or total nitrogen) of the extractable fraction, a rate of respiration per unit of cytoplasm possibly may be obtained. As can be seen from Table 12, the

TABLE 12

THE RATES OF RESPIRATION OF THE FRAGMENTS OF THE FROG GASTRULA (FIGURE 7) EXPRESSED IN VARIOUS TERMS

The extractable fraction is that part of the fragment which is extracted by 0.65 percent sodium chloride in a phosphate buffer at pH 7.4. The values express microliters of oxygen consumed per hour per unit of reference. From Sze, 1953c.

	FRAGMENT					
Reference	VEGE-TAL D	DORSAL LIP 1	ANIMAL POLE 2	ANIMAL POLE 3	VENTRAL LIP 4	VEGE-TAL V
Dry weight	1.2	3.7	4.4	3.9	2.7	0.90
Total nitrogen	12.0	37.0	53.0	46.0	30.0	8.2
Dipeptidase activity	55.0	28.0	19.0	23.0	25.0	33.0
Extractable fraction	(11)	12.0	13.0	13.0	13.0	(11)

ratios between the rates of respiration vary greatly depending upon the reference standard used for calculation of the rates. Rates calculated per unit of total nitrogen give values not very different from those calculated per unit of dry weight. Rates per unit of dipeptidase activity however entirely reverse the animal-vegetal gradient of respiration (3.9, 2.7, 0.9) to a vegetal-animal gradient (33, 25, 23). Finally, when the rates are computed per unit of extractable fraction the differences between the fragments of the gastrula become diminished and the difference between the dorsal lip (1) and the presumptive ventral lip (4) disappears. In comparing organizer and nonorganizer regions we must conclude that the former contains more extractable fraction (31 percent as compared with 21 percent) but that the respiration of that fraction is the same in both regions.

The frog gastrula then resembles the sea urchin egg in measurements of oxygen consumption. In both there are no differences detectable by manometric measurements. This does not exclude possible differences of oxygen consumption in the cortex of the eggs. Indeed evidence from investigations on the reduction of vital dyes points to differences in the rate of oxygen consumption.

THE SEARCH FOR QUALITATIVE DIFFERENCES IN RESPIRATION
OF THE ORGANIZER REGION

Coming back to the respiration of gastrula fragments we ask whether there are any differences in the kind of respiration in different parts. Ornstein and Gregg (1952) have investigated this problem by comparing the action of inhibitors on the respirations of the dorsal and ventral lips and also the anaerobic production of carbon dioxide of these regions. In another paper (Gregg and Ornstein, 1952) they report determinations of the amount of ammonia produced under anaerobic conditions. The method of dissection by Ornstein and Gregg is shown in Figure 8, while the results of their investigations are presented in Table 13.

From an inspection of Table 13 we find no significant differences in the behavior of the respiratory mechanisms of dorsal and ventral explants as affected by various chemical compounds. When a compound decreases the respiration of the dorsal lip it affects the presumptive ventral lip to the same extent. Thus as far as the kind of respiration in organizer versus nonorganizer regions is concerned, there are no detectable differences. The inhibition of gastrulation produced by many of these compounds in the entire gastrula, stage 10, is not correlated with a local action on the respiration of the dorsal lip, but is a general action on the respiration of the entire egg—at least of the dorsal and ventral fragments tested. Similar studies on the effect of various compounds on the respiration of the animal and vegetal halves of the sea urchin egg have revealed no measurable differences (Lindahl and Holter, 1940).

The data recorded in Table 13 correlate a morphological effect (inhibition of gastrulation) with (1) an inhibition of the cytochrome system by potassium cyanide, (2) an inhibition of the succinic dehydrogenase system by sodium malonate, (3) an inhibition of phosphorylation by sodium azide, (4) an inhibition of the enolase activity by sodium fluoride, and (5) an inhi-

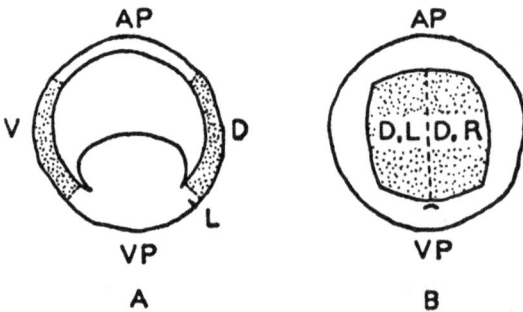

FIGURE 8. THE METHOD OF DISSECTION OF THE FROG GASTRULA TO OBTAIN COMPARABLE SAMPLES OF THE DORSAL AND VENTRAL LIPS FOR EXPERIMENTAL TREATMENT AND FOR CONTROLS
Dorsal and ventral fragments are dissected out. Then each is cut in half to give lateral halves. One lateral half is used for measuring the oxygen consumption in Holtfreter's solution, while the other lateral half is treated with a test solution and its oxygen consumption measured. AP = animal pole; VP = vegetal pole; D = dorsal side of gastrula; V = ventral side of gastrula; L = dorsal lip of blastopore; D,R = right half of dorsal fragment; D,L = left half of dorsal fragment. From Ornstein and Gregg, 1952.

bition of the sulfhydryl group by p-chloromercuribenzoic acid. Altogether this is an impressive array of reactions that appear to be necessary for the process of gastrulation and it supports the hypothesis of a carbohydrate metabolism similar to that of muscle. The seemingly anomalous effect of 2-4 dinitrophenol upon respiration does not speak against a correlation of respiration and gastrulation but rather for it. This compound is known to accelerate a respiratory metabolism which is not coupled with useful work and thus by diverting useful energy into a wasteful channel 2-4 dinitrophenol effects an inhibition.

In a more recent paper (Gregg and Ornstein, 1953), the authors studied the effects of inhibitors on morphogenetic movements in

TABLE 13

THE EFFECT OF VARIOUS CHEMICAL COMPOUNDS ON THE RESPIRATION OF DORSAL AND VENTRAL FRAGMENTS OF THE GASTRULA, STAGE 10, OF *Rana pipiens*

Dissections according to Figure 8; respiratory rates in microliters of oxygen per hour per mg. dry weight at 22° C. From Ornstein and Gregg, 1952.

NUMBER	SOLUTION	CONCEN-TRATION	MORPHOLOGICAL EFFECTS	RESPIRATORY RATES			INHI-BITION
				Dorsal Lip	Ventral Lip	RATIO	
1	Holtfreter's	100%	none	0.30	0.25	1.2	
2	Control	—	—	0.23	0.20	1.1	
3	Sodium azide	0.0005 M	inhibits gastrulation	0.08	0.06	1.4	65%
	Control	—	—	0.32	0.29	1.1	
4	Sodium azide	0.001 M	inhibits gastrulation	0.09	0.10	0.9	65%
	Control	—	—	0.31	0.27	1.1	
5	Potassium cyanide	0.0001 M	inhibits gastrulation	0.20	0.15	1.4	36%
	Control	—	—	0.26	0.18	1.4	
6	Potassium cyanide	0.0005 M	inhibits gastrulation	0.10	0.09	1.1	61%
	Control	—	—	0.32	0.26	1.2	
7	Sodium malonate	0.04 M	inhibits gastrulation	0.13	0.11	1.2	59%
	Control	—	—	0.36	0.26	1.4	
8	p-chloromercuribenzoic acid	0.0001 M	inhibits gastrulation	0.31	0.26	1.3	insignificant
	Control	—	—	0.28	0.23	1.2	
9	p-chloromercuribenzoic acid	0.001 M	inhibits gastrulation	0.10	0.08	1.2	64%
	Control	—	—	0.32	0.24	1.4	
10	Sodium fluoride	0.02 M	inhibits stage 14	0.19	0.14	1.3	40%
	Control	—	—	0.31	0.24	1.3	
	2,4, dinitrophenol	0.00001 M	inhibits gastrulation	0.66	0.53	1.2	110% stimulation

explants. Four types of movements were examined: 1, the embedding of gastrula endoderm into gastrular endoderm; 2, the stretching of the presumptive notochord on a base of gastrular endoderm; 3, the spreading of gastrular ectoderm over gastrular endoderm; and 4, the fusion of ectoderm with mesoderm when both are placed on a base of endoderm. The chart below summarizes their data. The effects of the inhibitors on the morpho-

CELLULAR PROCESSES EXAMINED IN EXPLANTS

Inhibitors made up in Holtfreter's solution, pH 8.0	1 Endoderm fuses into endoderm by formation of bottle-shaped cells	2 Notochord stretching on endoderm as during invagination	3 Ectoderm spreading over endoderm as during epiboly	4 Fusion of mesoderm with ectoderm on top of endoderm
2-4 dinitrophenol	normal	normal	inhibited	normal
95 N_2 : 5 CO_2	normal	normal	inhibited	normal
azide	normal	suppressed	suppressed	normal
p-chloromercuri-benzoic acid	normal	suppressed	suppressed	indeterminate
urethane	normal	suppressed	normal	normal
sodium barbital	suppressed	suppressed	suppressed	normal
sodium malonate 0.045 M	suppressed	suppressed	suppressed	suppressed
sodium chloride, 0.045 M	normal	normal	suppressed	normal

genetic movements are classified as normal when the movement occurs, inhibited when the movements are retarded, and suppressed when the movements fail to occur. It can be seen that the various inhibitors of the process of gastrulation of the whole gastrula act on different morphogenetic movements. Sodium malonate, for example, acts on all four types of movements, while urethane acts only on stretching of the notochord, and dinitrophenol on the spreading of the ectoderm. Such studies if pursued further may give information regarding the energy-yielding reactions for various morphogenetic movements.

Further investigations on a possible difference in kind of respiration have resulted in data on the ammonia and carbon dioxide liberation during anaerobiosis. The data reported by earlier investigators on anaerobic ammonia production by dorsal and ventral explants are not taken seriously by Gregg and Ornstein (1952), since the amount of ammonia produced is too low for accurate measurement. The authors conclude that only traces of ammonia are produced by frog gastrulae and that any difference between dorsal and ventral fragments has not been detected as yet.

Under anaerobic conditions the lactic acid produced liberates carbon dioxide from bicarbonates. The amount of carbon dioxide released probably is some function of the amount of lactic acid produced. In any case the rates of liberation of carbon dioxide from dorsal and ventral fragments of the frog egg are 0.11 and 0.10, respectively. The ratio of the rates, dorsal over ventral, is 1.1, and this ratio also is obtained when the rate of oxygen consumption of the dorsal fragment is divided by that of the ventral fragment. Thus the difference between the dorsal and ventral fragments is one of quantity and not of quality of respiration and glycolysis.

This conclusion is surprising to an experimental embryologist who has seen the differences in the developmental behavior of the cells from dorsal and ventral sides of the gastrula. How is it possible for two groups of cells to behave so differently while possessing the same respiratory metabolism? Either we accept the evidence or we contest it. If we accept it then we must look for differences at a different level of metabolism, namely the level at which the energy from respiration is utilized. If, for example, all the energy of respiration is transferred to adenosine triphosphate and reduced coenzyme, then we must search for different enzymes which catalyze a transfer of high-energy phosphate to different reactions and look for oxidation-reduction systems which may react with reduced coenzyme. This line of reasoning throws the whole emphasis on the acceptance of energy by different reactions by means of different enzymes. Either these different enzymes exist in all cells and are sorted out in some fashion or they are synthesized *de novo* in differentiating cells. If the latter is true, then a different mechanism for the synthesis of different enzymes in different cells must be postulated; and this brings us back to the same problem. Why, with the same respiratory mechanism, are different enzymes synthesized in different cells? We must recognize that fundamental differences pre-exist in different cells; otherwise we cannot obtain differentiation from an equipotential system.

What are the natures of these fundamental differences? If not respiratory then they may be differences in hydrogen ion concentration, or in other simple chemical compounds which when present in an egg result in a difference in the kind of chemical reactions which proceed within the egg. The determination of polarity of the *Fucus* egg by a difference in hydrogen ion concentration is a case in point.

HETEROGENEITY IN DISTRIBUTION OF FIVE VARIABLES
WITHIN THE GASTRULA

In any case let us continue the search for chemical differences between dorsal and ventral fragments of the frog gastrula. The distribution of five variables within the frog gastrula is recorded

TABLE 14

THE DISTRIBUTION OF CERTAIN VARIABLES IN THE GASTRULA,
STAGE 10, OF *Rana pipiens*

All values are in terms of dry weight of the fragment. Phosphate liberation means phosphate liberated from a brei at pH 3.5 by the action of phosphoprotein phosphatase on a phosphoprotein (Mezger-Freed, 1953). Rest of the data from Barth and Sze, 1953, and Sze, 1953c.

Variable	REGION OF GASTRULA						Ratio 3/V	Ratio 1/4
	D	1	2	3	4	V		
Respiration	1.2	3.7	4.4	3.9	2.7	0.9	4.3	1.3
Extractable fraction	8.7	31.0	35.0	32.0	21.0	8.7	3.6	1.5
Dipeptidase	0.022	0.131	0.231	0.177	0.113	0.027	6.3	1.2
Lipid	18.2	25.3	29.0	29.7	25.4	18.9	1.5	1.0
Total nitrogen	0.107	0.094	0.083	0.085	0.091	0.109	0.8	1.0
Phosphate liberation	27.0	27.0	21.0	21.0	26.0	24.0	0.9	1.0

in Table 14. The dissection of the fragments is the same as for studies on respiration, Figure 7 (Mezger-Freed, 1953; Barth and Sze, 1953; Sze, 1953c)

The regional differences in respiration are similar to those of the extractable fraction. The latter is a measure of the amount of cytoplasm in each fragment. There is roughly 3.6 times as much cytoplasm in the animal fragment (3) as compared with the vegetal fragment (V), while the respiration is 4.3 times higher in the animal fragment. As we have concluded previously, the rate of respiration of different fragments per unit of cytoplasm is about the same. The same situation does not prevail in the case of dipeptidase activity. The activity of the animal fragment is 6.3 times as much as the vegetal fragment. Calculated per unit of extractable fraction the dipeptidase activity of the animal fragment is still twice that of the vegetal fragment. If then the respiration is considered to be uniform throughout the gastrular cytoplasm the dipeptidase must be distributed in heterogeneous fashion. If, on the other hand, the dipeptidase activity is considered as homogeneously distributed, then the respiration cannot be so distributed. We must conclude that, whatever reference we use for calculating the distribution of variables in the gastrula, the cytoplasm is heterogeneous. Otherwise the animal-vegetal ratios of all cytoplasmic variables should be the same.

The distribution of lipid, total nitrogen, and the liberation of phosphate are special cases. Lipid is present in both the yolk and the cytoplasm and thus the animal-vegetal ratio is not so large as for respiration and dipeptidase activity. Similarly total nitrogen is made up of the nitrogen of the yolk and that of the cytoplasm, and since the yolk is mostly protein, the animal-vegetal ratio is less than one. The vegetal fragments contain more total nitrogen per unit of dry weight than do the animal fragments.

The phosphate liberated in a brei at pH 3.5 comes mainly from a phosphoprotein in the yolk (Mezger-Freed, 1953). An enzyme, phosphoprotein phosphatase (Harris, 1946), catalyzes the dephosphorylation of the protein. The vegetal fragments contain much more yolk than the animal fragments and thus much more phosphoprotein. It is not surprising then that the animal-vegetal ratio is less than one.

The difference between the organizer region and a non-organizer region is revealed by comparing fragment 1 with fragment 4. The ratio 1 : 4 is given in Table 14. Respiration, dipeptidase activity, and extractable fraction are present in higher amounts in the organizer region. The 1:4 ratio of 1.5 for extractable fraction means that there is more cytoplasm in the organizer region. Since respiration and dipeptidase activity are functions of the cytoplasm, it is only reasonable that the two should be higher in the organizer region. Since lipid and total nitrogen and phosphate liberation are functions of both yolk and cytoplasm, it is reasonable to find that they are present to an equal extent in both organizer and nonorganizer regions.

In general, as a result of the high concentration of yolk in the vegetal hemisphere any component of yolk, such as that revealed by total nitrogen or phosphoprotein, will necessarily be present in higher concentration in vegetal fragments. Similarly, as a result of the high concentration of cytoplasm in the animal hemisphere any variable which is a function of cytoplasm will be present in higher quantities in animal fragments. Since many measurements now are consistent with the hypothesis of a greater concentration of cytoplasm in the dorsal lip fragment, it is clear that the relations between dorsal and ventral fragments will be similar to the relation of animal to vegetal fragments. The difference between dorsal and ventral fragments in regard to any variable will be less than the difference between animal and vegetal fragments.

This greater concentration of cytoplasm on the dorsal side of the egg and of the gastrula is confirmed by Sze (1953c). Sze finds that the first segmentation furrow passes through the dorsal side of the egg more rapidly than through the ventral side. In the blastula and gastrula he finds that the cells on the dorsal side are smaller. Finally, Sze has made cell counts and finds a greater number of cells per unit volume on the dorsal side of the egg. These morphological observations combined with the fact that the extractable fraction (non-yolk fraction) is higher on the dorsal side make reasonably certain the conclusion of a higher concentration of cytoplasm on the dorsal side of the gastrula.

INVESTIGATIONS OF LOCAL DIFFERENCES WITHIN THE GASTRULAE
OF OTHER SPECIES

The results of investigations on other species are in agreement in some instances and disagree in others with the conclusions drawn above. The earlier work is treated completely and ably by Brachet (1950). We would like to call attention to some recent investigations by Gregg and Løvtrup (1950) on the gastrula, stage 10, of *Ambystoma mexicanum*. Their dissections are designed to give a more precise answer than have those used for the frog. Ten fragments of the gastrula are analyzed for a variable. They find the animal fragments to contain the greatest amount of, or exhibit highest activity for (1) extractable fraction (non-yolk nitrogen), (2) saponifiable fat, (3) total carbohydrate, (4) alanylglycine peptidase, (5) beta-glycerophosphatase, and (6) oxygen consumption. These variables are expressed in terms of total nitrogen, which is similar to dry weight and of course includes yolk. When referred to non-yolk nitrogen (extractable fraction) the animal fragments are highest in total carbohydrate, dipeptidase activity, and beta-glycerophosphatase activity.

Inspection of their data also shows that dorsal fragments (number 3 in their dissection and containing the organizer region) are higher than ventral fragments (number 9) in non-yolk nitrogen (extractable fraction), in total carbohydrate, in dipeptidase activity, and in beta-glycerophosphatase activity. If these differences are statistically significant then the dorsal fragments of *Ambystoma mexicanum* like those of *Rana pipiens* contain a greater amount of cytoplasm as compared with ventral fragments. And as a consequence, in part at least, the variables measured in dorsal fragments are higher than in ventral fragments.

An analysis of 7-minute-hydrolyzable phosphate (labile phosphate, probably adenosine triphosphate) of the dorsal and ventral fragments of the toad gastrula, *Bufo vulgaris*, was made by Fujii, Utida, Ohnishi, and Yanagisawa (1951). They found 66 micrograms of labile phosphate per 100 milligrams dry weight of the dorsal fragments. The ventral fragments contained only 46 micrograms. The amount of inorganic phosphate in dorsal and ventral fragments was about the same. Finally, their determinations of adenosine triphosphatase activity of dorsal and ventral fragments reveal that the activity of the former is 67 percent higher. The higher concentration of adenosine triphosphate and the higher ATPase activity of the dorsal lip region as compared with the ventral ectoderm are of such magnitude (from 40 to

67 percent) as to suggest a real difference between the cytoplasms of these two regions. For it is doubtful that the concentration of cytoplasm in the dorsal fragment is from 40 to 67 percent higher than in the corresponding ventral fragment. Measurements of extractable fraction and of other variables in the toad gastrula are needed to settle the question.

THE METABOLISM OF THE DORSAL LIP DURING INDUCTION

As far as local differences in the release of energy are concerned, the foregoing studies on the frog gastrula reveal no significant differences between dorsal and ventral fragments. Present indications therefore point to local differences at some other level in the chain of reactions connecting energy release with energy utilization. Perhaps the experimental embryologist might have predicted that the differences were not to be expected in the first reactions in the chain. For, knowing that the dorsal and ventral fragments are equipotential in development, such a wide difference as the occurrence of carbohydrate metabolism in one and protein metabolism in the other becomes difficult to imagine. To change from one type of metabolism to another would require a change in a large number of enzymes. The situation would be vastly simpler if the metabolisms were identical up to the last step in energy transfer, at which the presence or absence of a single enzyme might determine whether a notochord or ventral ectoderm differentiated.

Two studies of the metabolism of fragments of the frog dorsal lip during the process of induction have been made in our laboratory. Jaeger (1945) measured the glycogen content of the dorsal lip at the beginning of gastrulation and of the roof of the archenteron in the early neurula. A significant decrease in glycogen was found and this decrease possibly might be correlated either with the cellular movements during gastrulation, with the self-differentiation of the notochord and mesoderm of the dorsal lip, or with the process of embryonic induction of the neural plate. An experiment so designed that in one case the dorsal lip was combined with the presumptive epidermis, while in the second case the two were kept separate revealed that the glycogen content in the two cases was the same. Since induction of a neural plate resulted in the first case and no induction occurred in the second, the disappearance of glycogen during gastrulation did not correlate with embryonic induction.

A similar method was adopted by Barth and Sze (1952) to reveal any differences in the rates of oxygen consumption during embryonic induction of the neural plate. They report a slight increase in the rate of oxygen consumption of the preparations exhibiting embryonic induction over those in which no induction was possible.

The method of dissociating invagination and self-differentiation from the process of induction is open to criticism in special instances. For example, fragments of presumptive epidermis of *Ambystoma maculatum* will develop neural tubes and thus this species cannot be used for the dissociation of induction from other processes. In other species the isolated dorsal lip will form a neural tube and this result would be expected to complicate any interpretation of metabolic studies. Properly used, however, comparison of the "two-in-one" (combined organizer and epidermis) with the "two-in-two" (separated organizer and epidermis) preparations gives information regarding the process of neural induction.

We have deliberately omitted important histochemical studies in which metabolic differences in the intact gastrula have been measured because we have as yet nothing to contribute. We believe the histochemical studies to be as important as biochemical analysis, and they should accompany the latter whenever possible. The histochemical studies are reviewed by Brachet (1950).

With the tentative conclusion from the data presented in this chapter of identical modes of energy release in different regions of the egg, we turn to a consideration of the utilization of energy for a possible mechanism correlating metabolism with development. For the terminal links in energy transfer probably connect with the synthesis of enzymes or specific cytoplasmic proteins, either of which may differ in different parts of the gastrula. In the next chapter some aspects of protein metabolism will be discussed.

five

PROTEIN METABOLISM

CHANGES DURING DEVELOPMENT

From the foregoing chapters we have seen that carbohydrates furnish the energy for development and that the steps in metabolism of the frog egg are similar to those occurring in the metabolism of muscle and yeast. All the evidence, with the exception of a low respiratory quotient and traces of ammonia production, points to a steadily increasing carbohydrate combustion beginning possibly at fertilization, certainly present at the end of gastrulation, and continuing on through neurulation. Since most of the tissues and organs are determined by the late neurula stage we may conclude that chemical differentiation occurs at the expense of an oxidation of carbohydrates very probably coupled with phosphorylation, with adenosine triphosphate as an intermediary.

In addition, the metabolism of dorsal and ventral fragments of the gastrula appears to be identical, at least in the variables thus far measured. We turn then to the proteins for an investigation of the terminal steps in the energetics of development. The proteins, by directing or accepting the energy from a general, nonspecific carbohydrate metabolism might provide the mechanism for chemical differentiation. For it seems self-evident that the metabolisms of different parts of the egg cannot be identical in all the steps.

Let us begin our review of protein metabolism with a general picture of the distribution of nitrogen in the egg and the changes in nitrogen fractions with its development. A comprehensive study by Gregg and Ballentine (1946) reveals two important facts. First, only traces of ammonia and urea are present. Therefore only a trace of protein catabolism occurs, unless nitrogen is excreted, and a balance sheet shows no loss. Second, a separation of the protein nitrogen into a yolk-containing fraction and a yolk-free fraction shows that the former decreases with development while the latter increases.

TABLE 15

BALANCE SHEET FOR NITROGEN-CONTAINING COMPOUNDS IN THE *Rana pipiens* EGG

Adapted from Gregg and Ballentine, 1946. Some figures have been calculated from their graphs.

	MICROGRAMS IN	MICROGRAMS CHANGE IN		
	Fertilized Egg Stage 1 —jelly	Neurula Stage 14½ —jelly	Tailbud Stage 16½ —jelly —vitelline membrane	Hatched Embryo Stage 20 —jelly —vitelline membrane
1. Total nitrogen	162.0	−6.5	−13.0	−3.2
2. Extractable protein N	40.0	0.0	0.0	+2.8
2a Ultracentrifugible N	3.0	0.0	0.0	+0.2
3. Non-protein N	4.0	some loss	some loss	+1.0
4. Ammonia, urea, and glutamine N	0.61	—	+0.05	+0.07
5. Lipid N	3.0	—	—	—

Vitelline membrane N, stage 17 = 2 micrograms; fluid N, stage 17 = 6 micrograms; jelly N = 16 micrograms.

Studies on changes in protein nitrogen during development must be critically examined, as Table 15 compiled from the data of Gregg and Ballentine reveals. Three extracellular sources of nitrogen are present along with the nitrogen of the developing egg. The jelly contributes about 16 micrograms of nitrogen, the vitelline membrane, 2 micrograms, and the fluids of the egg (perivitelline, blastocoel, and archenteron), about 6 micrograms. Since the cytoplasmic protein nitrogen (extractable by 0.65 percent NaCl) weighs only 40 micrograms and the extracellular sources add up to 24 micrograms, a considerable error is introduced if these extracellular sources are ignored. For example, a determination of total nitrogen at fertilization and another after hatching might possibly show a loss of 24 micrograms without any actual change in the nitrogen of the egg proper.

If the jelly is removed before analyses are made a variable is introduced during development. Until about stage 13 to 14 the jelly may be removed without breaking the vitelline membrane. From stage 14 on there is increasing likelihood that in addition to removing the jelly (16 micrograms), the vitelline membrane (2 micrograms) and the combined fluids (6 micrograms) also are lost. For at about stage 13 the fluid of the archenteron escapes into the perivitelline fluid and, as Table 15 shows, the vitelline membrane plus the fluids within it yield 8 micrograms of nitrogen at stage 17.

At stage 20 hatching occurs and jelly, vitelline membrane, and fluid are lost; but the loss in total nitrogen does not appear to be as great at stage 20 (-3.2) as it was at stage 16½ (-13.0). It looks as if about 10 micrograms of nitrogen are transferred from the jelly and perivitelline fluid back into the embryo between stage 16½ and stage 20.

In any case the total nitrogen of the embryo decreases very slightly from about 160 micrograms at fertilization (162 minus 2 micrograms for the vitelline membrane) to 159 micrograms at stage 20, and measurements are not accurate enough to establish this decrease as certain.

Gregg and Ballentine made the very important distinction between the extractable protein nitrogen (presumably from cytoplasm) as opposed to the total protein nitrogen (cytoplasm + yolk and pigment granules). They obtained 40 micrograms of extractable protein nitrogen, and since only 3 micrograms of this came down in the form of granules at 20,000 revolutions per minute it is safe to conclude that only traces of yolk were present in their extractable fraction. The cytoplasmic protein nitrogen increases during development by about 2.8 micrograms and we conclude that cytoplasmic proteins are being synthesized at the expense of yolk proteins. There is also an increase in the nonprotein nitrogen which may possibly be correlated with the synthesis of nucleic acids. Graff and Barth (1938) reported that an increase in the percent of purine nitrogen of the total nitrogen occurred between fertilization and hatching. Since the total N changes very little if at all, some protein nitrogen probably is converted to purine nitrogen. Their graph shows an increase from 0.9 percent to about 1.9 percent. Taking a value of total nitrogen as 160 micrograms, this means 1.6 micrograms of purine nitrogen synthesis from some precursor. Gregg and Ballentine find an increase of 1.0 micrograms in the total nonprotein nitrogen.

Since the proteins do not appear to be lost by combustion or by excretion but do actually undergo transformations during development, our attention is directed toward the mechanism of the transformations. Are the yolk proteins broken down completely to amino acids which then are combined (with a great need of free energy) into the specific proteins of the differentiating cells; or are the yolk proteins merely converted by a simple modification (possibly requiring no free energy) to the cytoplasmic proteins?

The recent work of Friedberg and Eakin (1949) and also that of Eakin, Kutsky, and Berg (1951) give information on the reactivity of the proteins in the frog egg. In the first paper the activity of radioactive glycine was measured in the yolk proteins as distinguished from the supernatant (cytoplasmic) proteins. The activity of the yolk proteins in terms of counts per second per milligram of protein was 0.09, while that of the cytoplasmic proteins was 3.53. Thus the yolk proteins are not in a dynamic state of equilibrium with their amino acid components and may be considered relatively inert. The cytoplasmic proteins, as might be expected, readily incorporated radioactive glycine.

The extent of incorporation of radioactive glycine into the dorsal lip and ventral ectoderm of the gastrula was found to be different. The dorsal lip proteins registered an activity of 2.02 as compared with 1.04 for the ventral ectoderm. Since the yolk does not incorporate much glycine this difference is a cytoplasmic difference. Recall the greater amount of extractable fraction (Chapter 4) in the dorsal lip as compared with ventral ectoderm. If there is indeed a higher concentration of cytoplasm in the dorsal lip and radioactive glycine is incorporated chiefly by cytoplasm, then the results described are expected. It would be interesting to obtain the exact ratio of cytoplasms and activities of the dorsal and ventral fragments from the same gastrula

The second paper deals with the incorporation of radioactive methionine (S^{35}). Here the dorsal half (containing the dorsal lip) was compared with the ventral half (containing ventral ectoderm). The average activity of the proteins of the dorsal halves of *Rana pipiens* gastrulae was 57 percent, while that of the ventral halves was 43 percent. The dorsal half incorporated 32 percent more radioactive methionine than the ventral half. This is a very substantial difference in the activity of the proteins and again is probably in part a result of a difference in the amount of cytoplasmic proteins in the two halves. Whether this difference in the amount of cytoplasm in the two halves of the gastrula is great enough to account for the difference in activity of the proteins cannot be decided until determinations of the extractable fractions are made.

There are at least two different aspects of protein metabolism in the developing frog egg. One of these concerns phosphate metabolism in which the phosphoproteins provide the phosphate for the synthesis of nucleic acids (Kutsky, 1950). The other deals with the proteins as a source for the synthesis of enzymes, cytoplasmic proteins, and any possible morphogenetic proteins. The relation of the frog egg proteins to phosphate metabolism has been examined by the authors and is here presented in detailed form.

CHEMICAL AND ELECTROPHORETIC STUDIES ON FROG EGG PROTEINS[1]

Previous studies (Barth and Jaeger, 1950a, 1950b; Barth and Barth, 1951) have reported apyrase activity, phosphoprotein phosphatase activity, and some evidence suggesting a transfer of phosphate among the proteins found in a KCl extract of the frog egg or embryo. In addition a partial separation of some of the components of the extract was effected. Here we will report further progress in the separation of these components, as well as further studies on the properties of the extract. An electro-

[1] These investigations were made with the aid of a grant from the Committee on Growth acting for the American Cancer Society, and a grant (G-3322, c) from the division of research grants, Public Health Service, Department of Health, Education and Welfare, National Institutes of Health.

phoretic analysis of the extract was combined with chemical and enzymatic studies.

CHEMICAL PROPERTIES OF THE KCL EXTRACT. The crude KCl extract of the frog egg first is centrifuged at low speed to remove debris. It then is separated into three fractions by high-speed centrifugation: (1) a lipid fraction, which is discarded, (2) a clear yellow solution, which forms a precipitate when poured into 20 volumes of distilled water and is termed S in this paper, and finally (3) a black residue, R. R contains most of the enzymatic activity of the original crude extract, while S contains among other substances a phosphoprotein.

Properties of R. R contains enzymes which catalyze the release of phosphate from adenosine triphosphate, adenosine diphosphate, inorganic pyrophosphate, fructose diphosphate, and phosphoproteins. These enzymes are adsorbed on pigment granules at pH 6.8 but are eluted at pH 4.7. The phosphoprotein phosphatase is destroyed by boiling but is stable for at least a week at 3° C.

Properties of S. The composition of S has been studied by subjecting it to various treatments and then combining the resultant preparations with the enzymes of R. The following properties thus are revealed.

1. When S is combined with R the inorganic phosphate released comes from a phosphoprotein in which the bond between protein and phosphate can be broken by treatment with alkali (is alkali labile).

2. S plus sodium adenosine triphosphate combined with R results in a liberation of phosphate which is much less than that liberated from the combination of S and R as in (1). We have called this phosphate deficit "phosphate transfer" and have postulated an acceptor for that phosphate which does not appear as inorganic phosphate.

3. S plus sodium adenosine triphosphate form a complex such that the ATP is not removed by washing or dialysis. This "bound" ATP is responsible for the phosphate transfer of (2).

4. The phosphate deficit obtained in (2) divided by the expected phosphate as determined in (1) is defined as per-

centage transfer. The percent of phosphate transferred during a reaction is increased progressively by ions $Mg++$, $Ca++$, $Co++$, and $Mn++$. The greatest percent of transfer occurs with $Mn++$.

5. Mercuric chloride inhibits the liberation of phosphate from the KCl extract. $Mn++$ protects the extract from the action of $Hg++$. When ATP is added along with $Mn++$ and $Hg++$ an increased phosphate deficit results.

6. Treatment of S with relatively strong acid so alters its composition that upon incubation with R a phosphate deficit is no longer obtained in the presence of ATP. On the contrary, more phosphate is liberated in the presence of ATP. Thus normal S and acid-treated S respond very differently to the addition of ATP. However the phosphate surplus obtained with acid-treated S and ATP is converted to the usual phosphate deficit by the addition of $Mn++$.

The first two properties of S have been reported in previous papers (Barth and Jaeger, 1950a and b) and here we will concern ourselves with the additional chemical characteristics.

TABLE 16

BINDING OF ATP TO S

In each case the proteins were centrifuged down and washed before the 7-minute treatment with 1 N HCl at 100° C.

Experiment	Incubation Mixture	Bound ATP, 7 min. P $\mu g./100$ eggs
1	R + S + ATP, 30 min. incubation	149
	R + S − ATP, 30 min. incubation	10
2	R + ATP, 30 min. incubation	20
	S + ATP, 30 min. incubation	210
3	100% S + ATP	210
	50% S + ATP	100
	25% S + ATP	73
4	S 8 contains 94 $\mu g.$ alkali-labile P	94
	S 20 contains 146 $\mu g.$ alkali-labile P	141
5	S + ATP, 0° C., 15 min.	98

Binding of ATP to some Component of S. At the end of the period of reaction between R and S in the presence of ATP the proteins were centrifuged down and washed and treated with 1 N HCl at 100° C. for 7 minutes. Inorganic phosphate was split off during this period (Table 16, experiment 1), indicating that ATP had united with a protein. A test was made to see if the ATP united with the enzyme preparation, R, or with S. From Table 16, experiment 2, it is clear that S binds the ATP. With dilution of S the amount of bound ATP decreases (experiment 3). In experiment 4 it is seen that different stages of development yield differences in the amount of phosphoprotein extracted and similarly show differences in the binding of ATP, a fact which might be interpreted to mean that an ATP-phosphoprotein complex is formed. Finally, ATP combines with S at 0° (Table 16, experiment 5).

It is the bound ATP of the incubation mixture which is responsible for the phosphate deficit (Table 17). If R is similarly treated with ATP and combined with S no deficit occurs (Table 17, experiment 2). Thus the effect of ATP appears to be corre-

TABLE 17

BOUND ATP AND PHOSPHATE DEFICIT

In these experiments ATP was added to S at 0° C. for 15 min., after which S was spun down and washed to remove free ATP. This (S + ATP) complex then was combined with R and incubated under the usual conditions.

Experiment	Incubation Mixture	Difference ± ATP in Percent
1	S + R + ATP	51
	(S + ATP) + R	80
2	R(ATP treated) + S	0
	R(ATP treated) + (S + ATP)	80

lated with its reaction with S *and not to any effect on the enzyme preparation, R*. When ATP is added to S and the free ATP removed by washing, the resultant complex of S and ATP when combined with R yields deficits of 73 to 100 percent. This is more than the normal percent of deficit found in mixtures of S, R, and ATP. Thus the free ATP does not participate in the reaction. The increase in percent of deficit when S is allowed to stand with ATP may be a matter of a greater concentration of the complex (S-ATP) as compared with adding ATP to the reaction mixture and incubating immediately. If it is only the bound ATP which is responsible for the deficit, the latter would be greater where more ATP has been bound.

Metal Catalysis of the Phosphate Deficit. The percentage of phosphate which is not split during the reaction of S-ATP and R is somewhat variable in different preparations and a cofactor was suspected. The ions Mg++, Ca++, Co++, and Mn++ were added to the usual incubation mixture and the phosphate deficit was measured by the difference between incubation mixtures containing ATP and those lacking ATP. The metallic ions had little effect on phosphoprotein-phosphatase activity but did alter the amount of phosphate liberated in the presence of ATP (Table 18). The Mn++ ion produced the greatest phos-

TABLE 18

THE EFFECT OF METALLIC IONS ON PHOSPHOPROTEIN SPLITTING

The salts were 0.003 M in the incubation mixtures. The extracts were made from gastrula stages, and incubation was carried out in acetate buffer at pH 4.7, 27.5° C. for 30 minutes.

Experiment	Salt Added	Phosphate Liberated µg./100 eggs −ATP	+ATP	Difference ± ATP in Percent
1	None	237	170	−28
	Mg++	237	147	−38
	Ca++	268	147	−45
	Co++	243	107	−56
	Mn++	273	55	−80
2	None	249	187	−25
	Mg++	259	163	−37
	Ca++	275	147	−47
	Mn++	265	58	−78
3	Mg++	250	157	−37
	Co++	268	129	−52
	Mn++	270	52	−81

phate deficit. The values for phosphate deficit have not been corrected for ATPase activity, which goes on independently in incubation mixtures containing ATP. A direct analysis of 7-minute-labile phosphate before and after incubation showed that at least 40 units of the phosphate released in ATP-containing incubation mixtures comes from the splitting of ATP by R. The phosphate deficit thus is actually higher than the table shows, approaching in the case of Mn++, 100 percent "transfer."

When either R or S was treated with $Mn++$ and the free $Mn++$ washed out, no effect was obtained on the phosphate deficit during incubation. Thus under the conditions of the experiment, $Mn++$ did not form a complex with either the protein of S or of R. It is of course possible that longer exposure of S or R to $Mn++$ might have resulted in the binding of $Mn++$. It is also possible that a complex of S-Mn-R is formed on the addition of $Mn++$ to S plus R.

Effect of $HgCl_2$ and $MnCl_2$ on Phosphate Liberation in the Presence and Absence of ATP. The phosphate deficit obtained with ATP might result from the presence of small amounts of $Hg++$ in the ATP. Since $Mn++$ greatly increases the phosphate deficit caused by ATP, $Mn++$, and $Hg++$ in various combinations were tested on our KCl preparation (Table 19). Experiment 1 shows that less phosphate is liberated from phosphoprotein in the presence of $Hg++$. The addition of ATP produces an increased deficit at the lower concentrations of $Hg++$. The results from this experiment are consistent with an assumption that $Hg++$ in the ATP is responsible for the phosphate deficit.

The data in experiment 2, however, are not consistent with the above hypothesis. When $Mn++$ is added along with $Hg++$ there is no increased phosphate deficit such as is obtained with $Mn++$ and ATP. Rather, in the presence of $Mn++$ there is a decided increase in phosphate liberated in a concentration of 10^{-3} $Hg++$. If ATP contained $Hg++$ which was responsible for the phosphate deficit then the addition of $Mn++$ would be expected to decrease the phosphate deficit, because $Mn++$ decreases the inhibition produced by $Hg++$ in the absence of ATP. Compare experiment 1 with experiment 2 where 10^{-3} M $Hg++$ reduced the phosphate liberated from 249 to 77 micrograms, while 10^{-3} $Hg++$ + $Mn++$ reduced the phosphate only from 279 to 263 micrograms. The opposite effect of an increased phosphate deficit with $Mn++$ and ATP is actually obtained.

Experiment 3 summarizes the results. With no added metallic ions 340 micrograms of phosphate are liberated from a standard KCl preparation. With the addition of ATP the value drops to 264 micrograms, or a 23 percent deficit. With the addition of Mn^{++} a value of 372 micrograms is obtained and this drops to 87 micrograms in the presence of ATP, a 77 percent deficit. If Hg^{++} is added to the standard KCl preparation less phos-

TABLE 19

EFFECT OF Hg^{++} AND Mn^{++} ON PHOSPHATE LIBERATION IN NORMAL KCL EXTRACT

The total volume for incubation was 4 ml.: 2.5 ml. acetate buffer, containing R and S; 1.0 ml. of water or ATP; 0.25 ml. of water or 0.05 M $MnCl_2$; 0.25 ml. water or 10^{-5} to 10^{-2} molar $HgCl_2$.

Experiment	Incubation Mixture	PHOSPHATE LIBERATED $\mu g./100$ eggs		Difference ± ATP in Percent
		−ATP	+ATP	
1	R, S, H_2O	249	180	−28
	R, S, 10^{-5} Hg	238	170	−28
	R, S, 10^{-4} Hg	137	115	−16
	R, S, 10^{-3} Hg	77	111	+44
	R, S, H_2O	220	175	−20
	R, S, 10^{-5} Hg	245	157	−35
	R, S, 10^{-4} Hg	123	122	0
	R, S, 10^{-3} Hg	73	91	+25
2	R, S, Mn, H_2O	279	52	−81
	R, S, Mn, 10^{-5} Hg	310	55	−82
	R, S, Mn, 10^{-4} Hg	174	42	−76
	R, S, Mn, 10^{-3} Hg	263	24	−91
	R, S, Mn, H_2O	394	68	−82
	R, S, Mn, 10^{-5} Hg	361	81	−77
	R, S, Mn, 10^{-4} Hg	255	47	−82
	R, S, Mn, 10^{-3} Hg	371	—	—
	R, S, Mn, H_2O	213	—	—
	R, S, Mn, 10^{-4} Hg	123	29	−76
	R, S, Mn, 10^{-3} Hg	218	17	−90
	R, S, Mn, 10^{-2} Hg	46	26	−44
3	R, S, H_2O	340	264	−23
	R, S, Mn	372	87	−77
	R, S, 10^{-4} Hg	205	161	−21
	R, S, Mn, 10^{-4} Hg	297	62	−79

phate is liberated, but when Mn^{++} and Hg^{++} are added an increased amount of phosphate is obtained. Finally when Mn^{++}, Hg^{++}, and ATP are added a very large phosphate deficit of 79 percent occurs. It seems clear that the assumption of the presence of $Hg++$ in ATP cannot account for the action of ATP in producing a phosphate deficit.

The Properties of Acid-Treated S. During a somewhat lengthy attempt to fractionate S by precipitation at various pH's it was found that its properties were changed by treatment with relatively strong acid (0.16 N HCl). After such treatment and recovery by reprecipitation at pH 4.7, S was combined with R and some phosphate was liberated. When ATP was added, however, an increased amount of phosphate was obtained instead of the phosphate deficit usual with untreated S preparations. Referring to Table 20, experiment 1 shows that R as used in the following experiments has no ATPase activity. Experiment 2 shows that S(acid) has no ATPase activity or phosphoprotein-phosphatase activity. Experiment 3 compares S(acid) and S(normal) with regard to the effect of ATP. Although there is considerable variability in quantity, without exception the combination S(acid) + R results in more phosphate liberated when ATP is present, while S(normal) + R shows a phosphate deficit when ATP is present as compared to the same mixture without ATP.

Experiment 4 records a number of results in which the exact pH of precipitation after acid treatment and other variables were tested. In all cases more phosphate was liberated when ATP was present. Finally, experiment 5 demonstrates that $Mn++$ is able to prevent the excess phosphate from appearing and in addition produces a phosphate deficit such as is found with normal S, $Mn++$, and ATP.

When these results were obtained we concluded that acid treatment had two results: (1) an acquired ATPase cofactor, and (2) loss of a cofactor necessary for transfer. Neither S(acid) nor R alone gave high positive values in the presence of ATP, hence the necessity for postulating a cofactor for ATPase activity when the two were combined. The effect of $Mn++$ necessitated postulation of the cofactor for transfer, since if the only effect of $Mn++$ were to restore transfer, leaving the ATPase cofactor still active, the observed percent of transfer in $Mn++$ should

TABLE 20

EFFECT OF ATP AND MN + + ON PHOSPHATE LIBERATION FROM ACID-TREATED PHOSPHOPROTEIN FRACTION

S(acid) means that the phosphoprotein fraction, S, was treated with 0.16 N HCl for 15 min. It then was precipitated and combined with the enzyme, R. S(normal) means no treatment.

Experiment	Incubation Mixture	PHOSPHATE LIBERATED µg./100 eggs		Difference ± ATP µg./100 eggs
		−ATP	+ATP	
1	R	0	3	+3
	R	0	7	+7
	R	0	7	+7
	R	0	3	+3
2	S(acid)	0	0	0
	S(acid)	0	0	0
	S(acid)	0	0	0
3	S(acid), R	138	219	+81
	S(normal), R	256	209	−47
	S(acid), R	104	263	+159
	S(normal), R	370	304	−66
	S(acid), R	89	191	+102
	S(normal), R	363	305	−58
	S(acid), R	247	272	+25
	S(normal), R	318	238	−80
	S(acid), R	243	377	+134
	S(normal), R	340	177	−163
	S(acid), R	158	214	+56
	S(normal), R	373	188	−185
4	S(acid), R	330	460	+130
	S(acid), R	344	447	+103
	S(acid), R	241	393	+152
	S(acid), R	287	435	+148
	S(acid), R	374	478	+104
	S(acid), R	220	291	+71
	S(acid), R	136	196	+60
	S(acid), R	182	225	+43
5	S(acid), R, Mn	222	71	−151
	S(acid), R, Mn	133	60	−73
	S(acid), R, Mn	187	47	−140

be considerably lower with acid-treated S than with normal S. Such was not the case; S(acid) still responded to Mn++ by giving a large deficit (Table 20, experiment 5).

Summary of Properties of S. The foregoing experiments showed that the chemical composition of S is complex and the following substances were postulated in order to account for the results.

1. One or more phosphoproteins having a phosphate bond which is split by an enzyme in R.

2. An acceptor for the inorganic phosphate donated by the phosphoprotein.

3. A cofactor for ATPase activity which appears after acid treatment.

4. A cofactor for transfer of phosphate which is activated by Mn++.

While other alternate interpretations of the experiments were examined, we concluded that it was best to proceed as if at least the four postulated compounds had a real existence. Extensive experiments utilizing fractional dialysis and precipitation at various salt concentrations and pH's gave partial separation of S into fractions which behaved as if a separation of some of the postulated compounds had occurred.

Certain of the experimental results, however, were puzzling and prevented us from feeling secure about the system as postulated. The most annoying difficulty was our inability to separate the phosphoprotein donor from the phosphate acceptor despite a variety of different methods brought to bear upon this task. We *were* able to obtain preparations that had phosphoprotein-phosphatase activity alone, but failed to obtain from the residual S proteins another fraction which could be added to the phosphoprotein to restore "transfer" with ATP. The properties of donor and acceptor, furthermore, were suspiciously alike: the phosphate bond in both was alkali-labile, and it could be shown that the same enzyme preparation would split both donor and acceptor phosphate bonds.

The simplest explanation was, of course, that ATP inhibited phosphoprotein-phosphatase activity; that, for example, ATP and phosphoprotein were competing for the same enzyme, and that ATP had a greater affinity than the phosphoprotein for the enzyme. Yet the experimental results as a whole did not encourage this line of reasoning. Nor could we find any evidence that some impurity in our ATP preparations inhibited the enzyme, as has been shown. The effect that ATP exerted appeared to be through the S proteins and not to be a simple inhibition of or competition for the enzymes of R.

FRACTIONATION OF ACID-TREATED S AND INCUBATION TESTS WITH ENZYME ± ATP

We were obliged then, despite the failure of earlier attempts, to pursue the project of separating from whole S a fraction that was phosphoprotein donor and another fraction that could be established as the acceptor of phosphoprotein phosphorus through the intervention of ATP.

Fractional dialysis and fractional precipitation of the normal S proteins at different hydrion concentrations had not succeeded in yielding pure fractions. In the course of studies on acid-treated S, however, one fraction of it was found to precipitate at pH 8.5 to 9.5, while another fraction was obtained by precipitation at pH 2.1 to 4.5. The former was designated as S_1 and consisted of a voluminous yellow precipitate. The latter, termed S_2, comprised a much smaller fraction of the total and at pH 4.7 was of a rather waxy consistency and white color. When S_2 and S_1 separately were incubated with enzyme preparations most of the phosphoprotein was recovered in S_2, while S_1 although much greater in quantity contained much less phosphoprotein (Table 21).

When ATP was added to the incubation mixtures containing either S_1 or S_2 plus R, about the same amounts of phosphate were obtained as in the absence of ATP. If S_1 and S_2 were combined and added to the enzyme with and without ATP, then a surplus of phosphate was found with ATP. This surplus is similar to that obtained with whole acid-treated S. Since the surplus did not occur with either S_1 or S_2 alone, but only when both were combined, a two-factor system appeared to be in operation. When $Mn++$ was added to all incubation mixtures we found the usual high phosphate deficit similar to that obtained with whole acid-treated S.

We had obtained therefore two fractions of S that would produce the phosphate surplus when recombined with R and ATP, but had not as yet succeeded in reproducing with these

TABLE 21

PROPERTIES OF THE FRACTIONS OF ACID-TREATED S

S_1 = fraction precipitated at pH 8.5–9.5; S_2 = fraction precipitated at pH 2.5–4.5.

Experi-ment	Incubation Mixture	PHOSPHATE LIBERATED µg./100 eggs		Difference ± ATP µg./100 eggs
		$-ATP$	$+ATP$	
1	S(normal), R	268	135	-133
	S_1, R	64	65	$+1$
	S_1, S_2, R	280	425	$+145$
2	S_1, R	71	94	$+23$
	S_2, R	166	177	$+11$
3	S_1, R_2, R	372	486	$+114$
	S_1, R	122	110	-12
4	S_1, S_2, R	280	340	$+60$
	S_1, R	110	126	$+16$
	S_2, R	235	210	-25
5	S_2, R, Mn++	238	18	-220
6	S(normal), R	340	177	-163

fractions the phosphate deficit found in whole, untreated S. At this point we decided to try electrophoresis as a tool in the project of separating the two fractions. This proposed program depended upon the possibility that donor and acceptor might have different electrophoretic mobilities.

ELECTROPHORETIC STUDIES

ELECTROPHORETIC PATTERN OF THE KCL EXTRACT AND IDENTIFICATION OF TWO MAJOR COMPONENTS   The electrophoretic pattern of whole S was studied at stages ranging from stage 8 through stage 25+, the 12 mm. tadpole (Figure 9). Two distinct boundaries were obtained after 50 minutes at 20 m.a.

through a glycine buffer at pH 11.0 in a standard 10 ml. clinical cell of the Aminco Portable Electrophoresis Apparatus at 1.5° C. The slower moving boundary is a tall peak filling the screen at 70° magnification, while the faster moving boundary is a low

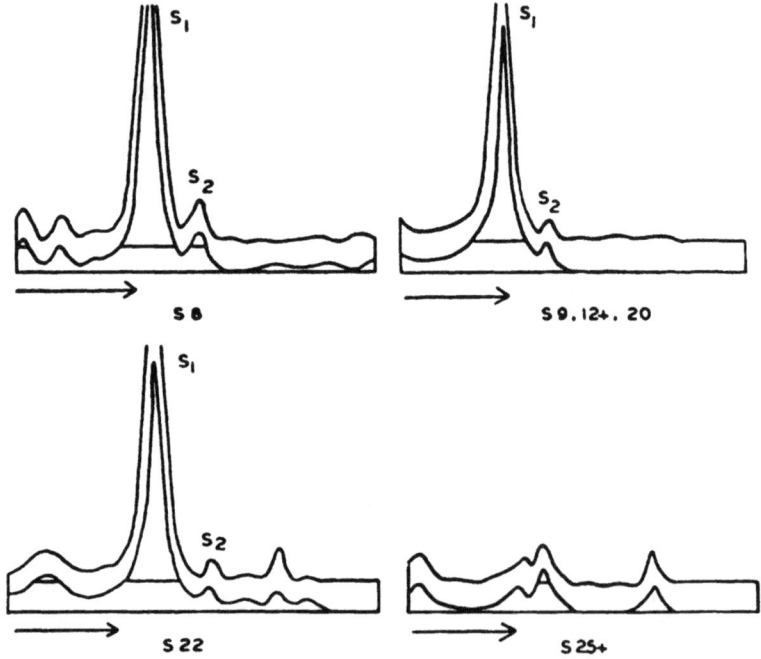

FIGURE 9. ELECTROPHORETIC DIAGRAMS OF PROTEINS CONTAINED IN THE S FRACTION OF KCL EXTRACTS AT DIFFERENT STAGES OF DEVELOPMENT

The tall peak present at stages 8 through 22 is designated S_1; the smaller component that moves ahead of S_1 is designated S_2. The diagrams are reproductions of tracings made at 70° magnification after 50 minutes migration at 20 m.a. through glycine buffer, pH 11 at 1.5° C. Ascending boundaries are presented. The S proteins had been dissolved in glycine buffer at pH 11, dialyzed overnight in the cold against glycine buffer, and centrifuged to obtain a clear solution for electrophoresis.

peak. The former is termed S_1, the latter, S_2. At stages 9, 12+, and 20 there is little change in the pattern, but at stage 22 peaks S_1 and S_2 are still present and a new very fast moving boundary appears. Finally, at stage 25+, S_1 and S_2 are very close to the base line and the new peak has increased slightly.

Since peaks S_1 and S_2 decrease in height at a time when the yolk is visibly reduced in quantity it was evident that one or the other contained the phosphoprotein and that the new fast-moving boundary represented a new compound synthesized at about stage 22.

A second study supplied evidence that the peak S_2 was caused by a phosphoprotein. Employing a large-scale fractionation of S into the two components earlier described, namely a

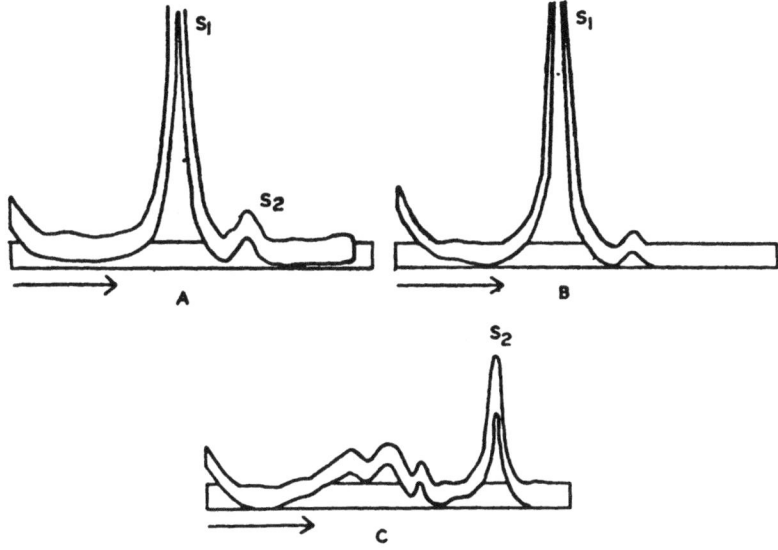

FIGURE 10. ELECTROPHORETIC DIAGRAMS OF YOLK PROTEINS IN KCL EXTRACT
A: Whole S fraction from KCl extract; B: Fraction precipitated at pH 9.5; C: Fraction precipitated at pH 2.1. Tracings of ascending boundaries were made at 70° magnification after 75 minutes migration through glycine buffer, pH 11, at 20 m.a. and 1.5° C.

fraction precipitated at a pH of 8.5 to 9.5 and another precipitated at pH 2.1 to 4.5, the electrophoretic pattern of each component was determined and compared with the original S. Figure 10 shows the patterns of the two components of S. The absolute heights of the peaks are not comparable from one diagram to another because different amounts of the proteins had to be used to yield convenient quantities for electrophoresis. The relative mobilities are comparable, however, and it is clear that the main component precipitated at pH 9.5 coincides with the tall S_1 peak observed in the whole KCl extract. The principal component of the fraction precipitated at pH 2.1 has a mobility much greater than that of S_1 and coincides more closely therefore with the S_2 peak of whole KCl extracts. Neither fraction is completely free of the other; thus a partial separation of the proteins had been achieved.

Aliquots of these two fractions used for electrophoresis were incubated with R. The results (Table 22) show that the phosphoprotein is chiefly in the fraction precipitated at pH 2.1 to 4.5, the fraction responsible for peak S_2.

TABLE 22

PROPERTIES OF FRACTIONS OF S USED FOR ELECTROPHORESIS

Aliquots of the two fractions used for electrophoresis (Figure 10) were incubated with R, and phosphate liberation was measured in the usual way. The values listed below have been corrected for concentration of protein present in each fraction as determined by measuring the areas under the two peaks S_1 and S_2 in a series of electrophoretic diagrams of normal KCl extracts. A conservative approximation of the relative amounts of S_1 and S_2 in a normal KCl extract indicates that S_1 exceeds S_2 by a factor of 10. That the smaller S_2 fraction contains practically all the phosphoprotein is apparent from the table below.

	$\mu g.$ P LIBERATED BY EQUIVALENT AMOUNTS OF FRACTIONS		
Incubation Mixture	$-ATP$	$+ATP$	$\mu g.$ Difference $\pm ATP$
S_1, R	7.1	12.7	$+5.6$
$S_1 + S_2$, R	156.8	169.6	$+12.8$
S_2, R	138.4	150.4	$+12.0$

Attempts to isolate a third fraction which appears in the electrophoretic pattern of gastrula extracts at 105 minutes have not as yet been successful. Some proteins do precipitate from pH 6.0 to 6.5, but so much phosphoprotein is present that the nature of the third component remains unknown. As can be seen from the pattern of S_2, several additional components are present. Some S_1 would be expected because of incomplete precipitation at pH 8.5 to 9.5, but in addition several other peaks are present. These are probably masked in the pattern of whole S by the large amount of S_1 present.

CORRELATION OF ELECTROPHORETIC PATTERN WITH THE BEHAVIOR OF COMPONENTS DURING INCUBATION   Having identified the phosphoprotein as peak S_2 in the electrophoretic diagrams given by a normal KCl extract, and recognizing the presence of the other main component, S_1, as the larger peak in such diagrams, we could now hope to follow the behavior of these two principal proteins during incubation with and without ATP. Would it be possible to recover the proteins at the end of an incubation during which "transfer" occurred and find in the electrophoretic pattern a new peak that would represent the acceptor-phosphate protein?

For these experiments large quantities of proteins were required and conditions that would produce a high percent of "transfer" were necessary. Ovarian eggs were used for these large-scale experiments and manganese was added to the incubation mixtures to promote a high percent of "transfer." Another condition for the success of this method of attack was of course the ability to recover all the proteins at the end of incubation and separate them into donor, acceptor, and dephosphorylated protein in order to observe any changes in the electrophoretic patterns that might result from incubation.

Our first attempt at recovery of the incubated proteins was based upon the observation that during incubation in the acetate buffer at pH 4.7 both S_1 and dephosphorylated S_2 appear to go into solution. The native phosphoprotein and part of S_1 remain insoluble in acetate buffer. At the end of incubation, therefore, the mixtures were centrifuged at 19,000 $\times$ g and the dephosphorylated proteins present in solution in the supernatant acetate buffer were made alkaline before dialyzing against glycine buffer at pH 11.5 for electrophoresis in the same glycine buffer. The residue from this centrifuging contained the remaining proteins, i.e., phosphoprotein and S_1 that had not been changed by incubation, as well as acceptor. These proteins were dissolved in glycine buffer at pH 11.5, the pigment centrifuged down, and the supernatant dialyzed preparatory to electrophoresis. Aliquots of the original incubation mixture were taken for the usual determination of phosphate released in presence and absence of ATP.

The results of an experiment in which a deficit of -50 percent was measured are shown in Figure 11 (A and B). The dephosphorylated protein shows the single tall, slow-moving peak previously identified as S_1 in whole extracts. The acceptor-phosphate pattern looks much the same as normal, unincubated S, containing in addition to the tall, slow-moving peak the smaller, faster moving peak previously identified with S_2. When these two fractions were mixed and submitted to electrophoresis the resulting pattern contained both peaks. We had to conclude therefore that the mobilities of native S_1, of dephosphorylated S_2, and of the hypothetical acceptor-phosphate were identical. All were fused in the single tall peak that lags behind the

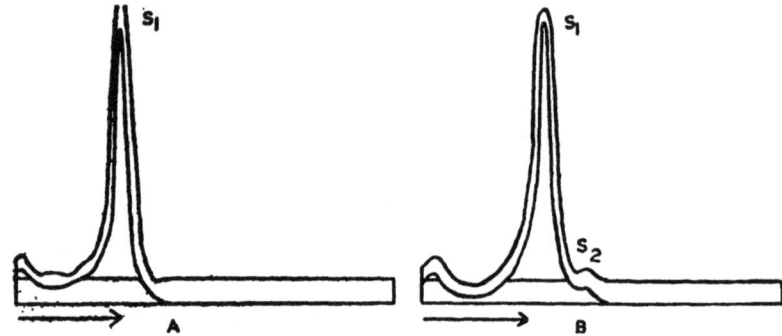

FIGURE 11. ELECTROPHORETIC DIAGRAMS OF S PROTEINS RECOVERED AFTER 30 MINUTES INCUBATION
A: Dephosphorylated protein; B: The pattern for acceptor and for dephosphorylated protein + acceptor, both of which gave identical patterns. The figures are reproductions of tracings of ascending boundaries made at 70° magnification after 50 minutes migration through glycine buffer, pH 11, at 20 m.a. and 1.5° C. See text for interpretation.

smaller, faster moving peak representing native S_2. Barring unknown side reactions, the results of this incubation study showed that the peak S_2 was caused by the phosphoprotein and when dephosphorylated the resulting protein had the same mobility as S_1. Of course it was also possible that the protein responsible for peak S_2 was modified in some other way by enzymes in R and its mobility altered.

It was possible that we had not recovered successfully all the proteins by the method employed and that this accounted for our failure to pick up a new peak representing the acceptor-phosphate. Another kind of incubation-electrophoresis experiment therefore was designed. A large quantity of S prepared from ovarian eggs was divided into two portions. Manganese and ATP were added to both, but one half was adjusted immediately to pH 11.5 by adding NaOH to dissolve the proteins and give the 0-minute situation. To the other half, R was added for incubation at 27.5° C. for 30 minutes. The incubation then was stopped and the proteins were dissolved by adding NaOH to pH 11.5. The pigment was centrifuged out of both mixtures, and the proteins dissolved in glycine at pH 11.5 in the supernatant solutions were dialyzed overnight and examined electrophoretically.

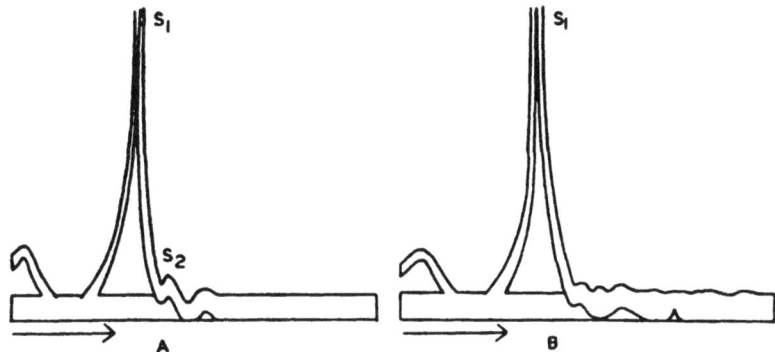

FIGURE 12. ELECTROPHORETIC DIAGRAMS OF S PROTEINS FROM TWO INCUBATION MIXTURES
A: S proteins present at 0 minutes incubation; B: S proteins present after 30 minutes incubation with R. The tracings of ascending boundaries were made at 70° magnification after 50 min. migration through glycine buffer at pH 11 at 20 m.a. and 1.5° C. Note that the height of the S_2 peak was reduced markedly as a result of incubation.

Figure 12 (A and B) shows the results. S alone at 0 minutes of incubation shows the tall, slow-moving S_1 peak and two smaller, faster moving peaks, the taller of which we recognize as S_2 on the basis of its relative mobility. The incubated proteins lack all but traces of the small, fast-moving components, but contain the usual tall S_1 peak. These data were interpreted provisionally as a visualization of the occurrence of transfer: we had started with native S_2. When it was split by the enzyme some of the phosphate presumably was transferred, while the remainder was simply released as free inorganic phosphate. But no longer was a peak present where native S_2 shows a peak, as would be expected had S_2 simply failed to break down in the presence of ATP, $Mn++$, and R. What this experiment lacked, however, was a parallel demonstration that transfer actually had occurred.

This fault was corrected in the next series of experiments where portions of the incubation mixtures were retained for determination of percent of transfer in the actual preparations

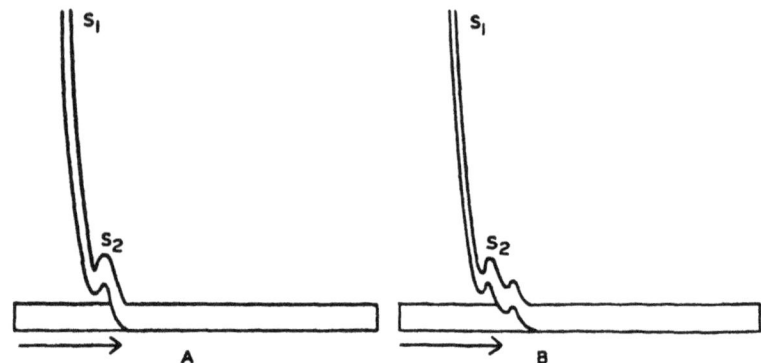

FIGURE 13. ELECTROPHORETIC DIAGRAMS OF S PROTEINS RECOVERED FROM INCUBATION MIXTURES

A: Control mixture at 0 minutes incubation; B: ATP-containing mixture after 30 minutes incubation. Note that the S_2 peak is almost as high at the end of 30 minutes incubation with ATP as it is before incubation. This suggests that much of the phosphoprotein does not break down in the ATP-containing mixture, a possibility that was confirmed by the parallel incubation test. The incubation tests for P liberation in small aliquots showed that 83.8 µg. P were split in controls lacking ATP, while only 30.3 µg. P were liberated in the presence of ATP, a phosphate deficit of −64%.

The incubation mixtures consisted of R and S obtained from 16 ml. of ovarian eggs. Each mixture contained 20 ml. of acetate buffer, pH 4.7, 8 ml. of 0.024 M ATP or H_2O, and 2 ml. of 0.2 M $MnCl_2$. From these large mixtures 2 ml. aliquots were removed for the determination of inorganic P at 0 and 30 minutes incubation.

The large control mixture was made alkaline to pH 11.5 to prevent phosphoprotein phosphatase activity and to dissolve the proteins, while the corresponding mixture containing ATP was incubated for 30 minutes at 27.5° C. before alkali was added. The enzyme-containing pigment fraction was centrifuged down at 19,000 g and the supernatant solution of S proteins was dialyzed overnight in the cold against glycine buffer at pH 11.5. The dialyzed protein solutions were centrifuged and the supernatants were examined in the electrophoresis apparatus.

to be studied by electrophoresis. The electrophoretic diagrams reproduced in Figures 13, 14, and 15 tell the story together with the corresponding data on "transfer" of aliquots taken from the same source.

A control incubation mixture, that is, one lacking ATP but containing all of S, R, and $MnCl_2$, gave the pattern shown in Figure 13A at 0 minutes of incubation. The other half of the same preparation incubated with ATP for 30 minutes showed at the end of this incubation, during which a 64 percent deficit was measured in a side test, a very similar pattern (Figure

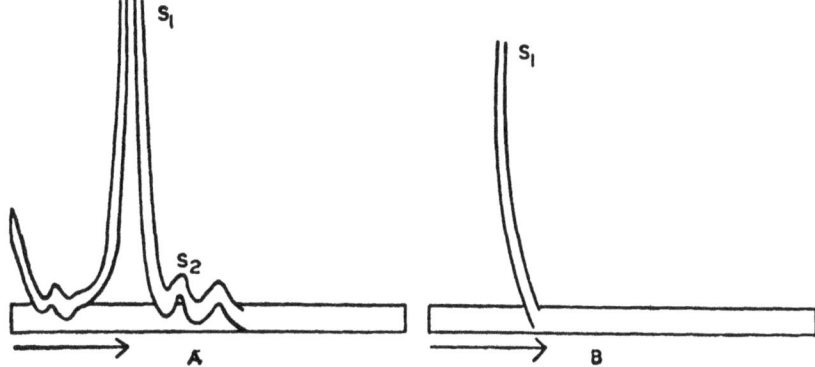

FIGURE 14. ELECTROPHORETIC PATTERNS OF S PROTEINS RECOVERED FROM INCUBATION MIXTURES
A: ATP-containing mixture at 30 minutes incubation; B: Control mixture at 30 minutes incubation. Note that the S_2 peak is lacking in B, where phosphoprotein has broken down in the absence of ATP. The incubation tests showed that 10.2 μg. P were split in an aliquot of the ATP-containing mixture, while an equivalent aliquot of the control gave 63.8 μg. P split during the incubation period. The P deficit ± ATP thus amounted to −85%.
The tracings reproduced above were made at 50 minutes migration and 70° magnification.

13B). If "transfer" had occurred in the ATP-containing mixture, the S_2 peak should have been reduced by about two thirds. The similarity of the patterns obtained from 0-minute controls and 30-minute incubation mixtures indicates that about two thirds of the protein simply did not break down in the presence of ATP.

Another experiment in which the incubation data showed an 85 percent deficit gave further evidence that very little phosphoprotein broke down in the presence of ATP and Mn + + and that "transfer" was simply failure of phosphoprotein breakdown (Figure 14). An experiment in which "transfer" was negligible served to control this conclusion (Figure 15). Transfer was low here because the concentrations of ATP and Mn++ were too low to give a large phosphate deficit. The pertinent point however is that when phosphoprotein did break down the char-

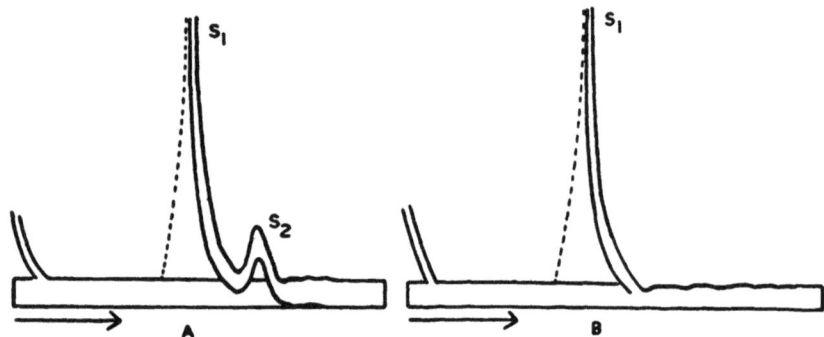

FIGURE 15. ELECTROPHORETIC DIAGRAMS OF S PROTEINS RECOVERED FROM INCUBATION MIXTURES
A: Control mixture at 0 minutes incubation; B: ATP-containing mixture after 60 minutes incubation.

P liberation in aliquots amounted to 111.5 µg. in the control and 91.3 µg. in the experimental mixture, a difference ± ATP of only −18%. Thus under the conditions of this experiment most of the phosphoprotein broke down during incubation and the S_2 peak correspondingly disappears from the electrophoretic pattern.

Tracings were made after 75 minutes migration under the usual conditions at a magnification of 70°.

acteristic S_2 peak disappeared from the electrophoretic pattern of the resulting protein extract (Figure 15B). When "transfer" occurred, as in Figure 14A, the S_2 peak remained at the end of incubation.

DIRECT DETERMINATION OF ALKALI-LABILE PHOSPHATE AFTER INCUBATION WITH AND WITHOUT ATP

The electrophoretic data were best interpreted as indicating a failure of native phosphoprotein to break down to the same extent when ATP is present as in the absence of this compound. A means of checking this interpretation appeared feasible in that aliquots from the incubation mixture at 0 minutes and at 30 minutes could be subjected to two simultaneous tests. The trichloracetic acid supernatants could be analyzed for inorganic phosphate released in the presence and in the absence of ATP, while the centrifuged residues from the TCA precipitation could be washed and hydrolyzed with 1 N NaOH for determination of alkali-labile phosphorus (phosphoprotein phosphorus). Such determinations of alkali-labile phosphorus had been made earlier in the study, but at that time we had concluded merely that the new phosphate acceptor must also be alkali-labile as was the phosphoprotein donor. The electrophoretic studies however cast a new light on the problem, in illustration of which we present the data of Table 23. An additional datum necessary to interpret

TABLE 23

EFFECT OF ATP ON ENZYMATIC SPLITTING OF PHOSPHOPROTEIN IN A NORMAL PREPARATION OF S

The inorganic P released during incubation in the presence of ATP comes in part from independent ATPase activity. Direct measurement of alkali-labile P gives the true splitting of phosphoprotein. Compare with Table 24 for the effect of ATP on splitting of S(acid).

	μg. P/100 EGGS		
	−ATP	+ATP	Difference ± ATP
1. Change in 7 min. P during incubation		+58	
2. Inorganic P released during incubation	440	365	−75
3. 2 corrected for 1	440	307	−133
4. Alkali-labile P split during incubation	470	360	−110

the incubation phosphate determinations when ATP is present is the amount of splitting of ATP by the enzyme ATPase also present in the mixture. This value must be subtracted from the apparent phosphate released by phosphoprotein-phosphatase activity, as is shown in the table.

The data summarized in Table 23 are consistent with the view that less phosphoprotein breakdown occurs in the presence of ATP, a conclusion confirmed and amplified by the more extensive series of experiments presented later.

The case for a phosphate acceptor was increasingly weakened by these findings, and our previous inability to separate phosphate donor from phosphate acceptor became more understandable. Another puzzling fact could now be interpreted tentatively. We had found that if a supply of "acceptor phosphate" were built up by permitting incubation to proceed in the presence of ATP, if the ATP then were washed out and fresh enzyme added, the "acceptor phosphate" was split by the enzyme. This now could be explained merely as the normal enzymatic breakdown of native phosphoprotein when ATP has been removed.

Although the phosphate deficit observed in the presence of ATP thus is best interpreted as an inhibition by ATP of phosphoprotein splitting, it must be stressed that the effect of ATP is upon the S proteins and not upon the enzyme fraction. The data cannot be interpreted either as inhibition of the enzyme by ATP or in terms of competition for the enzyme between ATP and phosphoprotein. Some singular effect of ATP upon the properties of one component of the S proteins was indicated, and this became our next point of attack. If ATP prevents splitting of phosphoprotein in normal preparations of S, to what may we attribute the *surplus* phosphate obtained when acid-treated S is incubated with enzyme and ATP? To answer this question and to probe further into the effects of ATP upon the S proteins the experiments described in the following sections were undertaken.

SOURCE OF THE SURPLUS PHOSPHATE WITH ACID-TREATED S AND ATP   The increase in inorganic phosphate which appeared in the presence of ATP was first attributed to a splitting of ATP and a cofactor for ATPase was suspected. However a direct analysis for ATP before and after incubation failed to substantiate this hypothesis. Some ATP did split to give inorganic phosphate, but the amount was not large enough to account for the surplus phosphate. Noting that the phosphoprotein-phosphatase activity after acid treatment was always lower than before such treatment, we had previously assumed a loss of phosphoprotein during precipitation. A direct analysis of the amount of phosphoprotein before and after precipitation showed that the loss was negligible and that we had as much alkali-labile phosphate after acid treatment as in our normal preparations. Thus the low values for phosphoprotein phosphatase were not a result of loss of the phosphoprotein but rather a partial inactivation of it by the acid treatment. In general only about half as much phosphate was released from acid-treated preparations as from the nontreated preparations, although each contained the same amount of alkali-labile phosphate.

We then tested the hypothesis that ATP acted on the acid-treated phosphoprotein to produce increased splitting. By measuring the alkali-labile phosphate before and after incubation with ATP and comparing this with the decrease in alkali-labile phosphate in the absence of ATP, we were able to show that ATP stimulates phosphoprotein splitting. The phosphate surplus obtained with ATP resulted from an increase in the liberation of phosphate from phosphoprotein (Table 24). The difference in amount of phosphate obtained with and without

TABLE 24

THE EFFECT OF ATP UPON ENZYMATIC LIBERATION OF P FROM PHOSPHOPROTEIN IN ACID-TREATED S

Experiment	Fractions Incubated	PHOSPHATE LIBERATED DURING INCUBATION µg./100 eggs			Change in Alkali-labile P, Difference ± ATP µg./100 eggs
		−ATP	+ATP	Difference	
1	S(acid), R	155	454	+299	+235
2	S(acid), R	140	282	+142	+102
3	S(acid), R	230	433	+203	+104

ATP in the incubation mixture is always more than the difference in the amount of alkali-labile phosphate split with and without ATP. The discrepancy between the two differences is made up by the phosphate liberated from ATP by the independent ATPase activity of R.

ANALYSIS OF THE EFFECT OF DENATURATION OF S AND OF THE ROLE OF ATP

Why does acid treatment of S prevent the phosphoprotein from breaking down enzymatically to the same extent as untreated S? Since both acid-treated S and untreated S are in the form of a precipitate at pH 4.7, it was possible that a difference in physical state of the phosphoprotein might account for the differences in amount of phosphate liberated by the action of the enzyme. Accordingly acid-treated S was redissolved in the original KCl buffer and reprecipitated by a 20-volume dilution. This should have resulted in the same physical state as untreated S. The acid-treated phosphoprotein, however, still failed to break down to the same extent as untreated phosphoprotein (Table 25, experiment 3).

The effect of acid treatment upon S appeared to resemble a true denaturation of the proteins. The effect of heat denaturation upon the splitting of S was examined therefore. Denaturing S in a water bath at 100° C. for 10 minutes so changed the proteins that very little phosphate was liberated by the enzyme (Table 25, experiment 1). However, the addition of ATP resulted in a liberation of more phosphate, as was the case with

TABLE 25

THE EFFECT OF HEAT DENATURATION OF S ON ENZYMATIC SPLITTING OF PHOSPHOPROTEIN, AND THE RESULTS OF ATTEMPTED REVERSAL OF DENATURATION

Experiment	Preparation	P RELEASED DURING INCUBATION µg./100 eggs			ALKALI-LABILE P SPLIT DURING INCUBATION µg./100 eggs		
		$-ATP$	$+ATP$	$\Delta \pm ATP$	$-ATP$	$+ATP$	$\Delta \pm ATP$
1	S(boiled)	29	223	+194	65	275	+210
	S(normal)	438	312	−126	550	340	−210
2	S(boiled)	70	354	+284			
	S(boiled), redissolved in alkali, and reprecipitated in acid	110	388	+278			
3	S(acid), redissolved in alkali	105	220	+115			
	S(acid), redissolved in KCl and reprecipitated in water	102	209	+107			

acid-treated S. These conclusions were confirmed by direct measurement of alkali-labile (phosphoprotein) P split during incubation.

It was clear that both heat denaturation and acid denaturation resulted in a decrease in the amount of phosphate liberated. Furthermore ATP appeared to reverse the effects of this denaturation, if we can draw this conclusion from the fact that

ATP brought about an increase in the liberation of phosphate. Attempts to reverse the denaturation by dissolving the proteins at pH 11.5 and reprecipitating at pH 4.7 were unsuccessful in that the phosphoprotein still failed to break down enzymatically (Table 25, experiments 2 and 3).

THE EFFECT OF CONCENTRATION OF ATP ON THE LIBERATION OF PHOSPHATE FROM PHOSPHOPROTEIN, NORMAL AND DENATURED

The results thus far indicated that while ATP has an inhibiting effect on the liberation of phosphate from the normal phospho-

TABLE 26

THE EFFECT OF CONCENTRATION OF ATP ON PHOSPHOPROTEIN SPLITTING IN NORMAL AND ACID-TREATED PREPARATIONS OF S OBTAINED FROM OVARIAN EGGS

The concentrations of ATP represent mg. of sodium adenosine triphosphate per ml. contained in the 1 ml. added for incubation to 2.5 ml. of acetate buffer, pH 4.7, with S and R. All values represent µg./100 eggs.

Incubation Mixture	Concentration of ATP mg./ml.	P LIBERATED DURING INCUBATION			ALKALI-LABILE P SPLIT	P from ATP
		−ATP	+ATP	Difference ±ATP	Difference ±ATP	
S(normal), R	2	476	381	−95	−113	+18
S(normal), R	4	476	325	−151	−198	+47
S(normal), R	8	476	455	−21	−79	+58
S(acid), R	2	210	173	−37	−28	—
S(acid), R	4	210	311	+101	+67	+34
S(acid), R	8	210	461	+251	+200	+51

protein, it stimulates splitting of the acid-treated, denatured phosphoprotein. In some experiments with acid-treated phosphoprotein we noticed an inhibitory effect of ATP when the concentration of ATP in the incubation mixture was lowered. A more extensive series of experiments showed that both inhibition and stimulation of splitting of the phosphoprotein may occur when the concentration of ATP is varied. Table 26 and Figure 16 show the effect of various concentrations of ATP on the amount of phosphate liberated from S(normal) and S(acid). Although there was actually more alkali-labile phosphate in S(acid), much less phosphate was liberated as compared with that formed during incubation of S(normal). Figure 16 shows that the addition of 1 mg./cc. of ATP produced a decrease in the amount of phosphate liberated by S(acid) and S(normal). With increase of the concentration of ATP to 4 mg./cc. a sharp difference in the behavior of S(acid) and S(normal) was revealed. The amount of phosphate split from S(normal) decreased, while that split from S(acid) increased. Further increase in concentration of ATP to 8 mg./cc. produced an increase in the amount of phosphate liberated both from S(normal) and S(acid). An analysis of the alkali-labile phosphate at each concentration of ATP showed that the variations in amount of phosphate liberated resulted from variations in the amount of phosphoprotein that split off phosphate. The variations were not caused by differences in ATPase activity, which increased as a linear function of ATP concentration.

These results as presented in Figure 16 show that ATP in low concentrations inhibits the liberation of phosphate from phosphoprotein in acid-treated S but in high concentrations ATP stimulates release of phosphate. The effect of ATP on normal S differs in that while inhibition of phosphoprotein splitting occurs at lower concentrations the effect of the higher concentration of ATP is merely to increase phosphoprotein splitting to a level almost equal to that measured in the complete absence of ATP.

THE EFFECT OF S_1 AND OF ATP ON THE SOLUBILITY OF PHOSPHOPROTEIN   In view of the new data on the effect of ATP on phosphoprotein splitting, some puzzling visual observations of changes that occurred during incubation became susceptible to further investigation. This new point of view led also to a successful separation of relatively pure S_1 and S_2 fractions, which in turn permitted more detailed analysis of the precise effects of ATP upon the proteins contained in S.

During incubation of whole S with the enzyme in the absence of ATP the mixture always became clear and black. The same preparation *with* ATP remained conspicuously gray and turbid, and had to be stirred frequently to prevent the proteins from settling out. In the former control mixtures lacking ATP we had

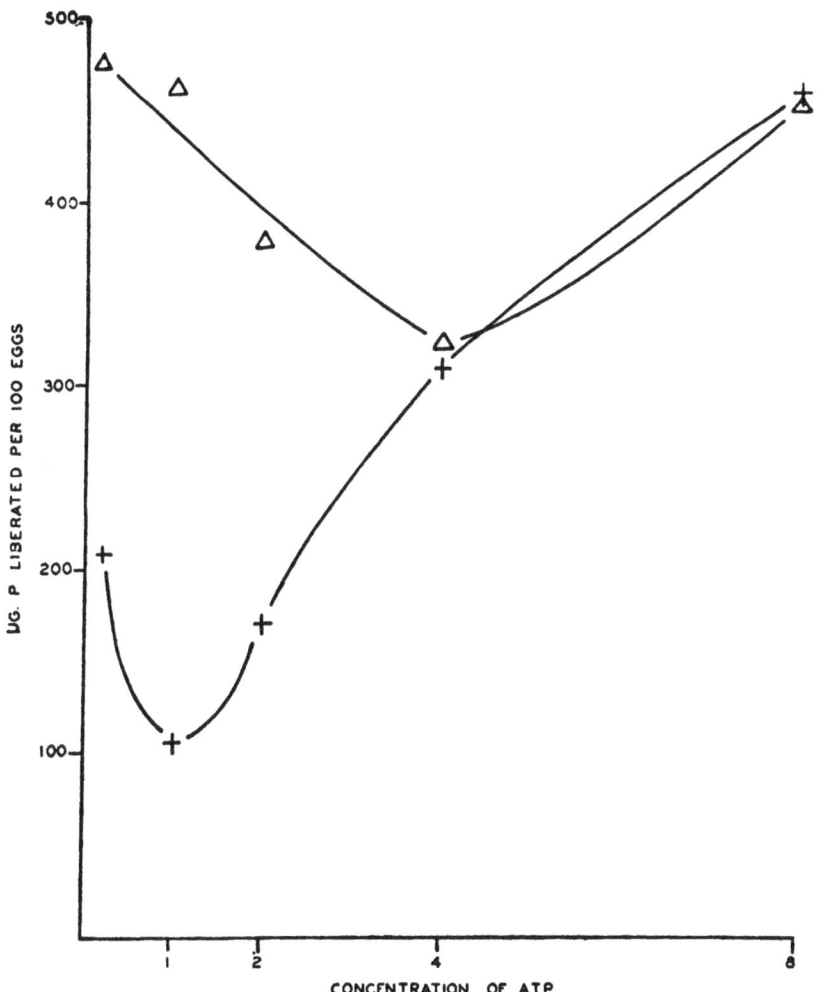

FIGURE 16. THE EFFECT OF CONCENTRATION OF ATP ON PHOS-
PHOPROTEIN SPLITTING IN NORMAL AND ACID-TREATED S
The concentrations of ATP represent mg. ATP per ml. contained in the
1 ml. added for incubation to 2.5 ml. of acetate buffer, pH 4.7, with S
and R. The curve drawn through triangles represents values for S(normal); the curve marked with + signs represents values for S(acid).

observed furthermore that both fractions of S became progressively soluble in acetate buffer as incubation proceeded. Electrophoretic diagrams of the proteins that centrifuged down in acetate buffer showed a progressive decrease in height of both peaks as incubation proceeded from 0 to 30 minutes, while the acetate supernatant gave increasing amounts of TCA-precipitable material.

Since the significant change during incubation is splitting of phosphoprotein, this suggested that native S_2 is insoluble in acetate buffer but becomes soluble upon dephosphorylation. It followed then either that S_1 also became soluble through some change during incubation, or that S_1 if it were free of S_2 was soluble in acetate buffer at pH 4.7 to begin with.

TABLE 27

THE PHOSPHOPROTEIN CONTENT OF TWO FRACTIONS OF S

The fractions were separated on the basis of solubility in acetate buffer, pH 4.7, as explained in the text, and were incubated with R in the usual way. The values for P liberated in absence of ATP show that most of the phosphoprotein appears in the acetate-insoluble S_2 fraction.

Fraction	P LIBERATED DURING INCUBATION µg./100 eggs		
	$-ATP$	$+ATP$	Difference ± ATP
S_1	44	70	+26
S_2	159	194	+35

If S_1 and S_2 could be obtained as pure fractions without prior acid treatment this hypothesis could be tested and the mutual effects of the two fractions with respect to solubility could be examined independently of the effects of incubation. A new method of purifying S_1 and S_2 was devised. Whole S precipitates were dissolved in glycine buffer at pH 11.5. When the pH was lowered to 8.8 by addition of HCl, crude S_1 precipitated and was collected by centrifugation. This S_1 could be freed of its S_2 traces by dissolving the crude S_1 in acetate buffer at pH 4.7 and centrifuging out the insoluble S_2 precipitate.

The S_2 fraction had to be recovered from the original supernatant obtained at pH 8.8 by adding enough HCl to give pH 3.0 to 3.5, at which point S_2 precipitates. The crude precipitate was taken up in acetate buffer at pH 4.7 to dissolve remaining traces of S_1 and after centrifugation the precipitate was designated S_2.

The efficacy of this method of fractionation was tested by incubating the fractions with R. The results, shown in Table 27, show that 78 percent of the phosphoprotein is recovered in the acetate-insoluble S_2 fraction. The low plus values obtained with ATP are attributed to independent ATPase activity.

The solubility properties of S_1 and S_2 obtained in this manner may be tabulated as shown in Table 28. It is seen that either S_2 or ATP will precipitate S_1 from solution in acetate buffer, pH 4.7. As extracted from the egg, S_2 plus S_1 form a complex that is insoluble in acetate buffer, pH 4.7. When the enzyme is

TABLE 28

SOLUBILITY OF FRACTIONS OF S AT DIFFERENT HYDROGEN ION CONCENTRATIONS

Fraction	pH		
	11.5	8.0–9.5	4.7–1.6
S_1	soluble	insoluble	soluble
S_2	soluble	soluble	insoluble
$S_1 + S_2$	soluble	—	insoluble
S_1 + ATP	—	—	insoluble

added, its phosphoprotein-phosphatase activity removes S_2, leaving S_1 free to dissolve. When ATP is present, less S_2 is split and less S_1 becomes soluble. Most of the proteins remain in the S_1-S_2 insoluble complex.

THE ROLES OF S_1 AND S_2 IN THE ATP-DENATURATION EFFECT With the method described above for obtaining native S_1 and S_2 as relatively pure fractions, it became possible to analyze the effects of denaturation and of ATP upon the enzymatic liberation of phosphate from phosphoprotein.

Some of the data on these problems are tabulated in Table 29. If S_1 is first removed from S_2, denaturation of S_2 has little effect on enzymatic splitting of the phosphoprotein, and ATP has, if anything, an inhibiting effect on splitting, as in normal, whole, untreated S (Table 29, experiment 1). Experiment 2 shows the efficacy of the fractionation method, since S_1 is almost free of phosphoprotein, and the plus value obtained with ATP is due to ATPase activity at this high concentration of ATP (8 mg./cc.). The second line of experiment 2 shows that if S_1 and S_2 are separately denatured by boiling each fraction before recombination for incubation with the enzyme, S_2 still splits enzymatically to the same extent as does normal, untreated native S, and ATP has no effect on the splitting of S_2. From experiment 3, however, it is seen that if S_1 and S_2 are mixed and well homogenized together before heat denaturation, subsequent incubation with the enzyme results in relatively high plus values in the presence of ATP. That is, ATP now is necessary for phosphoprotein breakdown of this denatured S_1-S_2 complex, as was the case for boiled or acid-treated *whole* S.

TABLE 29

THE EFFECT OF ATP ON THE SPLITTING OF DENATURED
PHOSPHOPROTEIN ALONE AND IN COMBINATION
WITH THE S_1 FRACTION

Ovarian eggs were used and the data have been converted to μg. P split by fractions obtained from approximately 100 eggs. See text for full explanation.

Experi-	COMPONENTS OF S IN THE INCUBATION MIXTURE				
ment	Native	Denatured	−ATP	+ATP	Difference ± ATP
1	S_1		53	55	+2
	S_1	S_2	171	153	−18
		S_2	260	215	−45
2		S_1	10	46	+36
		S_1, S_2	141	150	+9
		S_2	151	165	+14
3		S_1	4	36	+32
		$(S_1 + S_2)$	34	122	+88
		S_2	109	128	+19

This suggested that the effect of ATP upon phosphoprotein splitting was somehow brought about through the intervention of S_1. This conclusion was given further weight by the following experiment. Native S_2 was isolated without prior acid treatment and incubated with various concentrations of ATP. A control without ATP gave the amount of P split from S_2 without ATP. The resulting curve shown in Figure 17 proves that ATP has no effect on the splitting of S_2. The slight rise with concentration of ATP is attributed to increase in ATPase activity with concentration of ATP. When this figure is compared with Figure 16 where whole S(normal) and S(acid) were incubated with various concentrations of ATP, it becomes obvious that the plus and minus values given by S(normal) and S(acid) depend upon the complex between S_1 and S_2.

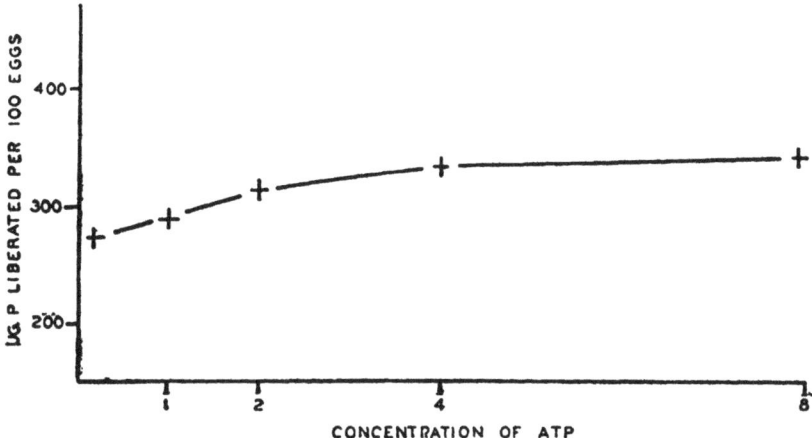

FIGURE 17. THE EFFECT OF CONCENTRATION OF ATP ON THE SPLITTING OF NATIVE PHOSPHOPROTEIN ISOLATED FROM ITS COMPLEX WITH S_1

The concentrations of ATP represent mg. ATP per ml. contained in the 1 ml. added for incubation to 2.5 ml. of acetate buffer, pH 4.7, with S_2 and R. Compare this curve with those presented in Figure 16, where S_2 was not isolated from the complex.

DISCUSSION

Although interpretation of the foregoing data cannot be entirely definitive at the present time, some new information about the yolk proteins of the frog egg has emerged from the study. That ATP *could* play a role in the control of phosphoprotein breakdown in the developing embryo is apparent, and the beginnings of an insight into the nature of this control mechanism are presented in the above sections. It seems probable, at the present state of our information, that ATP may not act as a direct intermediary in the transfer of phosphate from phosphoprotein donor to acceptor, but rather that its function may reside in a control of the amount and locus of phosphoprotein breakdown.

It seems clear also that if our test-tube situation holds for the egg, ATP does not act directly upon the phosphoprotein but rather through the intervention of an important protein component of "yolk" not previously described in the literature. The component, designated S_1, is by far the more prominent of the two main components of our yolk extracts, both in the electrophoretic diagrams and when judged in terms of volume of precipitate. Although S_1 is not a phosphoprotein, i.e., contains no alkali-labile phosphate, it appears to be associated with true phosphoprotein in a rather firm complex such that it goes into solution along with S_2 when this substance is dephosphorylated by phosphoprotein-phosphatase activity. Furthermore both proteins disappear at the same stage of development. ATP has no effect on the enzymatic hydrolysis of native S_2 alone when the phosphoprotein is removed from its complex with S_1 without denaturing either protein. But if the natural state in the egg of the two proteins is represented by the extracted S_1-S_2 complex, then ATP very definitely can be visualized as a controlling factor in the breakdown of phosphoprotein in the embryo.

Although the point was not investigated exhaustively, it is fairly certain that the S_1 and S_2 proteins are derived from yolk, as defined and extracted by Panijel. A series of experiments not reported earlier in this chapter was run on washed yolk platelets obtained by Panijel's (1951) method and dissolved and fractionated according to our usual procedure. The results obtained when such fractions from washed yolk were combined with yolk-free cytoplasmic enzyme preparations approximated the data obtained by our usual procedure on KCl extracts of whole eggs.

It was surprising to find that such a large bulk of the yolk proteins consisted of a component that was not phosphoprotein. A previous study of egg proteins obtained from the yolk of amphibian eggs was reported by Lawrence, Miall, Needham, and Shen in 1943. These workers separated the total euglobulin fraction into three main components characterized by different salt concentrations at which they precipitated, by different P/N ratios, and different viscosity properties. Two of these fractions contained a high ratio of P to N, and the third a lower proportion of P; in other words, no one of their fractions was free of phosphoprotein. The fractions obtained by the English workers cannot be compared directly to ours, since such different methods were used, but it is apparent that yolk contains more than one protein.

The significance of the present work depends, of course, upon its applicability to the actual utilization of yolk within the developing embryo. Phosphoprotein followed throughout development according to the Schneider fractionation (Kutsky, 1950) shows a slight decrease during early development. Our own data indicate that there can be little loss in phosphoprotein until after hatching. These findings relate to overall utilization, however, and it is possible that small local utilizations have not been picked up by methods employed heretofore. We can go on, therefore, to speculate upon the possible role of ATP in yolk utilization.

The phosphoprotein is for the most part (92 percent) localized in the yolk granules as an insoluble complex with a protein S_1. As such it is probably not available to the enzyme phosphoprotein phosphatase. The small amount (8 percent) in the cytoplasm may be presumed to be available to the enzyme and the control of the splitting of phosphoprotein depends upon pH, state of denaturation, and the concentration of ATP. With the pH of the unfertilized egg at 7.0 and a progressive increase to pH 8.0 at gastrulation (Dorfman, cited in Brachet, 1950), the phosphoprotein-phosphatase activity must be at a minimum. In the tailbud stage the pH drops to 7.0 once more but even at this value phosphoprotein-phosphatase activity is very low.

Conditions in the egg thus are such that only a slow liberation of phosphate can occur. The rate of liberation of phosphate is about the same as its utilization for synthesis of nucleic acids and other constituents, since the concentration of the acid-soluble fraction of the frog's egg does not change appreciably from fertilization to hatching. Such a situation might be expected if ATP controls the liberation of phosphate in the egg as it does in test tube. From Figure 16 it is seen that as the concentration of ATP is increased the rate of liberation of phosphate is decreased. This relationship holds between 0 and 4 along the axis representing concentration of ATP. A calculation of the actual concentration of labile phosphate of the egg from Barth and Jaeger (1947) gives a value of 0.0004 molar. This concentration lies between 1 and 2 on the graph of Figure 16. The actual concentration of labile phosphate in the egg is therefore in the range where an increase in ATP would result in a decrease in liberation of phosphate from phosphoprotein. Conversely, should the concentration of ATP drop, the amount of phosphate liberated from phosphoprotein would increase. Thus a rather fine adjustment of the control of the concentration of inorganic phosphate is provided by the concentration of ATP. The concentration of ATP would in turn be a function of the general phosphate metabolism of the egg.

The protein metabolism may be correlated with phosphate metabolism by combining some of the data in this chapter with the results obtained by Kutsky and others. The stored proteins exist as an insoluble complex of S_1 and a phosphoprotein, PS_2, as shown in the diagram below. Kutsky (1950) reported the liberation of phosphate from PS_2 to the extent of 7 percent of the total P. The proteins S_1 and S_2 become soluble, and Gregg and Ballentine (1946) measured an increase in soluble proteins. Kutsky further concluded that the P liberated became incorporated into labile phosphates and finally became nucleic acid P. We have shown that the concentration of ATP regulates the splitting of phosphate from phosphoprotein in the test tube. Cohen (1953) found that glycolysis was of the phosphorylating type and that inorganic P was necessary for glycogen utilization in homogenates. Thus the phosphoprotein, PS_2, liberates inorganic P, which is incorporated by glycolysis into ATP. As the concentration of ATP momentarily increases, the splitting of PS_2 is inhibited. When the concentration of ATP decreases momentarily by transfer of P to a nucleic acid precursor the splitting of PS_2 is increased. This mechanism insures a steady source of phosphate from PS_2 and as a by-product the yolk proteins become soluble and thus available for synthesis into cytoplasmic constituents.

A self-regulating mechanism involving the concentration of ATP, the concentration of inorganic P, and the liberation of phosphate from phosphoprotein may provide the explanation for the constancy of the concentration of labile phosphates and inorganic phosphates found by Barth and Jaeger (1947). One hitherto unexplained fact reported in their paper may become reasonable now. They found an increase in inorganic phosphate and a decrease in labile P under anaerobic conditions. But the

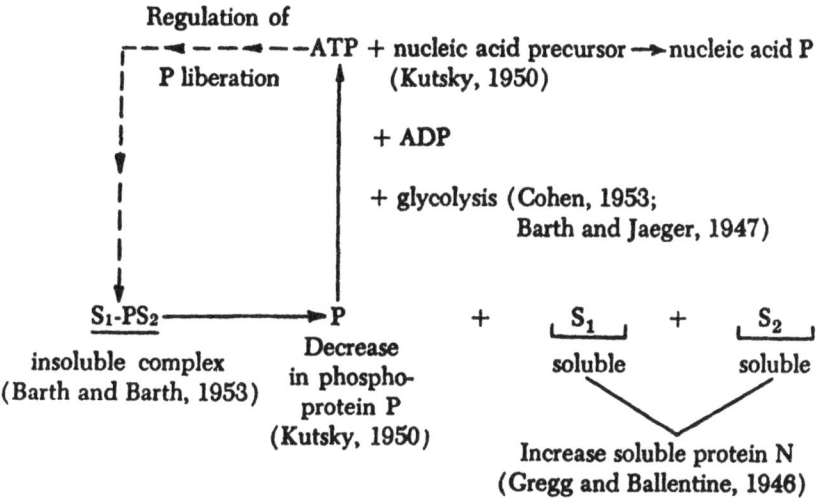

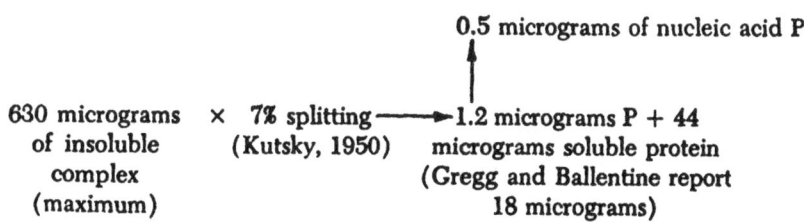

increase in inorganic P was greater than the decrease in labile P, indicating another source of inorganic P. Since the concentration of ATP decreases under anaerobic conditions an increased splitting of phosphoprotein should occur and the concentration of inorganic P should rise. The extra source of inorganic P during anaerobiosis probably is phosphoprotein.

A second method by which ATP could control the concentration of inorganic phosphate derived from phosphoprotein is by means of reversing the effects of denaturation. If proteins in the cell are partially denatured by surface denaturation as in the case of catalase (Kaplan, 1952), then the denatured phosphoprotein-protein complex would liberate phosphate very slowly (Figure 16). A local increase in the concentration of ATP such as occurs in the dorsal lip (Fuji, Utida, Ohnishi, and Yanagisawa, 1951) might result in a stimulation of phosphoprotein splitting and more inorganic phosphate for synthesis of nucleic acid and other constituents of protoplasm. Since we know nothing about the state of denaturation of the phosphoprotein-protein complex within the egg further speculation is useless. We wish to emphasize, however, that ATP by either controlling or stimulating phosphoprotein breakdown may play an important role in the metabolism of the egg as a whole and in the localized metabolism of its parts.

METHODS

The proteins used for this study were obtained as described in previous papers. The ionic strength of the buffer used for extraction was found to be critical in some types of experiments. In all our experiments we used a solution of 0.7 M with respect to KCl and 0.34 M with respect to sodium succinate. The volume of the eggs together with some dilute Ringer's solution (1/10 Ringer's minus carbonate and phosphate) was 2 ml. The buffer was added to make a total volume of 10 ml. Thus the final concentration for extraction was always 4/5 that of the original buffer, i.e., 0.56 M KCl and 0.27 M sodium succinate.

The object of the study was in no case the quantitative determination of proteins. Rather the aim was to analyze the composition of the extract and to examine the behavior of its components with respect to one another. The raw data were converted for publication as follows: When embryos were used the data were expressed as μg. P released by a stated combination of components obtained from 100 eggs. When ovaries were used the data were converted to μg. P released by an estimated 100 eggs, the estimate being based upon the observation that 1 ml. of fresh ovary contains approximately 100 eggs.

Since the experimental methods varied in detail depending upon the question to be answered by any given experiment, we will present the basic procedure for two kinds of experiments. *All operations preceding actual incubation were carried out at 0° to 3° C.*

I. Procedure for obtaining components of the KCl extract and for determining enzymatic liberation of P by various combinations of these components.

A. Extraction: Transfer 320 embryos to a graduated centrifuge tube and adjust volume to 4 ml. with 0.1 Ringer's solution that lacks phosphate and carbonate. Transfer eggs plus Ringer's solution to a ten Broeck tissue grinder and add 16 cc. of 0.7 M KCl in succinate buffer at pH 6.8. Homogenize thoroughly, divide homogenate between two heavy-duty 12 ml. centrifuge tubes, stopper and clamp tubes to a rotating wheel for 30 minutes. Centrifuge for 12 min. at 1,000 g. Decant gray supernatant liquid from both tubes into clean tube and mix.

B. Separation of R and S: Divide crude extract among four small lusteroid tubes and centrifuge 15 min. at 19,000 g. Remove yellow supernatant solutions by drawing each up carefully into a capillary-tipped pipette attached to a hypodermic syringe barrel, using care to avoid disturbing the black residue and the floating yellow fat cap. Transfer each of the four supernatants to a clean lusteroid tube and recentrifuge for 15 min. at 19,000 g to complete removal of fat and of insoluble residue.

To the packed down black residues obtained in the first high-speed centrifuging gently add about 4 ml. of 0.6 M KCl in succinate pH 6.8. Scrub caked fat from sides of tubes with a glass rod and discard the cloudy fluid. Invert tubes and keep cold till the S fraction has been worked up.

Using the capillary-syringe pipette, remove the clear, defatted yellow supernatants resulting from the second high-speed centrifuging. Pool and mix. Pipette 2 ml. aliquots into 8 lusteroid tubes containing 40 ml. each of ice-cold distilled water. Stir. Centrifuge 15 min. at 1,000 g. Discard supernatant liquids.

C. Acid treatment of S proteins: To the 8 precipitates of S add a little at a time a total of 20 ml. weak veronal buffer, pH 8.5 to 8.7 (made by diluting Longsworth's buffer with water in the proportions 1:10). Scrub out precipitates from bottom of tubes with a rounded glass rod. Pool the suspensions in a ten Broeck homogenizer, grind, and transfer to a 40 ml. lusteroid tube.

Add a small drop of phenolphthalein and 2.0 ml. of 2 N HCl. Mix and let stand 15 minutes. Neutralize to faint pink with 2.5 N NaOH and adjust pH to 4.7 with HCl. Centrifuge down the flocculent precipitate at 1,000 g for 15 min.

D. Preparation of incubation mixtures: R and S have been obtained as precipitates as described in sections B and C above. If S(normal) is to be used, section C is omitted; if S(acid) is to be tested, the procedure outlined in C is followed. In either case the procedure and quantities of reagents used for incubation are henceforth the same.

To each S precipitate add 2.5 ml. acetate buffer, pH 4.7. Swirl tube to loosen precipitate and pour with quick motion into ten Broeck grinder. Add 0.5 ml. extra acetate buffer per each 5 ml. suspension of S to allow for drainage loss. Homogenize.

To each R precipitate add in two separate additions a total of 5.5 ml. S-acetate suspension, scrubbing the sticky black R precipitate from its tube with a glass rod. Decant the R, S-acetate suspension to a small ten Broeck grinder and homogenize.

Of this mixture 2.5 ml. aliquots are removed to prearranged pairs of 12 ml. heavy-duty pyrex centrifuge tubes. To one member of a pair add 1 ml. ATP (4 mg./ml. $Na_4 \cdot ATP \cdot 3H_2O$, Rohm and Haas); to the other, 1 ml. water. Insert a 1 ml. blow-out pipette graduated to 0.01 ml. in each tube and mix by blowing.

E. Incubation of S proteins with enzyme $\pm$ATP: Take a 0.75 ml. aliquot from each member of each pair ($\pm$ATP) of mixtures. Add 0.25 ml. 75 percent TCA to each sample and stir with glass rod. Remove rack of incubation mixtures to 27.5° C. waterbath. Allow tubes to reach temperature before starting timer. Stir mixtures every 7 min. At 15 min. and at 30 min. remove 0.75 ml. aliquots from each mixture, adding 0.25 ml. 75 percent TCA as above. Remove all rods from tubes and let stand 6 minutes in cold. Centrifuge 7 min. at 1,000 g.

F. Determination of inorganic phosphate in TCA supernatants: The remaining steps are carried out at room temperature. Decant each supernate into 0.5 ml. 2.5 N NaOH in 10 ml. Klett tube. This neutralizes the strong TCA. The TCA residues may be retained if desired for determination of alkali-labile phosphate.

To each neutralized supernate add in turn 1.2 ml. of acid + molybdate and 0.4 ml. Fiske-Subbarow reductant. Make to 10 ml. with water, mix and time for 10 min. color development. Read with a no. 66 filter in a Klett colorimeter. A blank and duplicate 20 μg. standard P samples are carried through simultaneously.

II. Procedure for electrophoresis of S proteins.

A. Extraction and removal of R from S proteins: Preliminary trials showed that large quantities of extract were required to yield sufficient phosphoprotein, (S_2), for a clear electrophoretic diagram. Ovarian eggs therefore sometimes were used as source material.

Remove ovaries from female; cut into small pieces; wash in 0.1

Ringer's solution. Transfer 2 ml. volume of clumps of eggs into each of 8 heavy duty pyrex 12 ml. centrifuge tubes. Homogenize briefly each in turn with 8 ml. 0.7 M KCl, succinate pH 6.8, in ten Broeck. Return to centrifuge tubes, stopper, and put on rotor for 30 min. extraction. Centrifuge 12 min. at 1,000 g. Pool and mix the supernatant extracts. Ovarian tissue thus is discarded with the residue, the homogenizing having been too brief to break small cells, and the extract is composed of components from the large, easily ruptured ovarian eggs.

Centrifuge twice at high speed as described in Section IB, removing supernatant solutions of S with a capillary pipette to defat. This gives 8 tubes of black, enzyme-containing R and a total of about 60 ml. of clear S solution.

Precipitate aliquots of S in 20 volumes of water and centrifuge, as described in Section IB.

B. Acid treatment and separation of S_1 and S_2: Take up whole S precipitates in total of 80 ml. of 1/10 concentration Longsworth veronal buffer, pH 8.6. Homogenize in ten Broeck and pool. Add 8 ml. 2 N HCl plus 12 very small drops of phenolphthalein. Let stand in cold 15 min. Add 6.5 ml. of 2.5 N NaOH and about 8.5 ml. of 0.1 N NaOH till the pH is 9.8. Centrifuge 25 min. at 1,000 g. The precipitate contains most of S_1, while S_2 remains in solution at this alkaline pH.

To supernate add 0.8 ml. of 2 N HCl and adjust pH to 2.8. Centrifuge 15 min. at 1,000 g. The precipitate contains S_2. Take up precipitate in 40 ml. 1/10 concentration Longsworth buffer, homogenize in ten Broeck, add phenolphthalein and NaOH to pH 9.3. This should precipitate any S_1 impurity and leave S_2 in solution. After centrifuging 15 min. at 1,000 g. add to supernate 0.65 ml. of 2 N HCl to pH 2.1 and centrifuge again. The precipitate is used as the S_2 fraction.

Purify the crude S_1 precipitate obtained as a precipitate at pH 9.8 as follows: Make up a solution containing 60 ml. dilute Longsworth buffer, 6 ml. of 2 N HCl, 4.5 ml. of 2.5 N NaOH and enough additional weak alkali to bring the pH to 9.6. Homogenize S_1 precipitate in this mixture. Centrifuge 25 min. at 1,000 g. Repeat this washing. The precipitate contains S_1 that is relatively free of S_2.

C. Preparation of S_1 and S_2 for electrophoresis: To the bulky, yellow precipitate of S_1 add about 70 ml. of glycine buffer, pH 11.5. Homogenize to dissolve. Adjust pH with NaOH to pH 11.5. Take an aliquot of about 12 ml. into a cellophane dialysis bag and dialyze in an Aminco mechanical dialyzer against 300 ml. of glycine buffer, pH 11.5 in cold for 18 hours.

Note that only a fraction of the S_1 yield from the original extract of ovarian eggs is used for electrophoresis. The eggs contain so much of this protein that the total yield is too great for electrophoresis.

To the relatively meager, waxy, white precipitate of S_2 obtained in Section II, B, add 12 ml. of glycine buffer at pH 11.5. Homogenize to dissolve. Dialyze in Aminco mechanical dialyzer against 300 ml. of glycine buffer for 18 hours in cold.

Remove S_1 and S_2 from dialysis sacs and centrifuge 15 min. at 19,000 g. The clear solutions then are ready for transfer to the electrophoresis cell. The buffer solution against which the proteins were dialyzed is used for electrophoresis. Small samples of both protein solution and outside buffer solution are retained for checking the pH at the end of dialysis.

D. Electrophoresis: The standard 10 ml. clinical cell of the Aminco Portable Electrophoresis Apparatus is used, and a current of 20 m.a. At pH 11.5 the proteins migrate from the negative toward the positive electrode and the ascending boundary is observed on the screen to move from left to right.

Because we wished to observe the boundaries at frequent intervals throughout electrophoresis, simple pencil tracings on transparent paper were made by placing small sheets of the paper over the screen and tracing around the outlines of the peaks and baseline. A magnification of 70° was found convenient. Tracings were made at 0, 15, 30, 50, 75, 90, and sometimes 105 min., by which time the ascending boundaries had completed their left to right migration. The cell then was cleaned and used for a second electrophoresis the same day.

LITERATURE CITED

Barnes, M. R. 1944. The metabolism of the developing *Rana pipiens* as revealed by specific inhibitors. J. Exp. Zool., 95: 399–418.

Barth, L. G., and L. Jaeger. 1947. Phosphorylation in the frog's egg. Physiol. Zool., 20: 133–146.

——— 1950a. The role of adenosine triphosphate in phosphate transfer from yolk to other proteins in the developing frog egg. I. General properties of the transfer system as a whole. J. Cell. and Comp. Physiol., 35: 413–436.

——— 1950b. The role of adenosine triphosphate in phosphate transfer from yolk to other proteins in the developing frog egg. II. Separation of the system into component enzymes, phosphate donor and phosphate acceptor. J. Cell. Comp. Physiol., 35: 437–460.

Barth, L. G., and Lucena J. Barth. 1951. The relation of adenosine triphosphate to yolk utilization in the frog's egg. J. Exp. Zool., 116: 99–122.

Barth, L. G., and L. C. Sze. 1952. The organizer and respiration in *Rana pipiens*. Exp. Cell. Research, 2: 608–614.

——— 1953. Regional chemical differences in the frog gastrula. Physiol. Zool., 26: 205–211.

Brachet, J. 1950. Chemical Embryology. Interscience Publishers, Inc. New York. 533 pp.

Cohen, Adolph I. 1953. Studies on glycolysis during the early development of the *Rana pipiens* embryo. In press.

Eakin, Richard M., Phyllis B. Kutsky, and William E. Berg. 1951. Protein metabolism of amphibian embryo. III. Incorporation of methionine into protein of gastrulae. Proc. Soc. Exp. Biol. Med., 78: 502–504.

Friedberg, Felix, and Richard M. Eakin. 1949. Studies in protein metabolism of the amphibian embryo. I. Uptake of radioactive glycine. J. Exp. Zool., 110: 33–46.

Fujii, T., S. Utida, T. Ohnishi, and T. Yanagisawa. 1951. The Apyrase activity and adenosinetriphosphate content of the organizer region of *Bufo vulgaris formosus*. Annotationes Zoologicae Japonenses, 24: 115–119.

Graff, Samuel and L. G. Barth. 1938. Cold Spring Harbor Symposia, 6: 103–108.

Grant, Philip. 1953. Phosphate metabolism during oögenesis in *Rana temporaria*. J. Exp. Zool., 124: 513–544.

Gregg, John R. 1948. Carbohydrate metabolism of normal and hybrid amphibian embryos. J. Exp. Zool., 109: 119–134.

Gregg, John R., and Robert Ballentine. 1946. Nitrogen metabolism of *Rana pipiens* during embryonic development. J. Exp. Zool., 103: 143–168.

Gregg, John R., and Søren Løvtrup. 1950. Biochemical gradients in the axolotl gastrula. Compt. Rend. Lab. Carlsberg, Sér. Chim., 27: 307–324.

Gregg, John R., and Norma Ornstein. 1952. Anaerobic ammonia production by amphibian gastrulae explants. Biol. Bull., 102: 22–24.

──── 1953. Explant systems and the reactions of gastrulating amphibians to metabolic poisons. Biol. Bull., 105: 466–476.

Harris, D. L. 1946. Phosphoprotein phosphatase, a new enzyme from the frog egg. J. Biol. Chem., 165: 541–550.

Healy, Eugene. 1952. Metabolism of the frog hybrid *Rana pipiens* × *R. clamitans*. Doctoral Thesis, Columbia University Library.

Jaeger, L. 1945. Glycogen utilization by the amphibian gastrula in relation to invagination and induction. J. Cell. and Comp. Physiol., 25: 97–120.

Kaplan, J. Gordin. 1952. The biological activity and physical state of intracellular catalase. Physiol. Zool., 25: 123–131.

Kutsky, Phyllis B. 1950. Phosphate metabolism in the early development of *Rana pipiens*. J. Exp. Zool., 115: 429–460.

Lawrence, A. S. C., Margaret Miall, Joseph Needham, and Shih-Chang Shen. 1943. Studies on the anomalous viscosity and flow-birefringence of protein solutions. II. On dilute solutions of proteins from embryonic and other tissues. Jour. Gen. Physiol., 27: 233–271.

Lindahl, P. E., and H. Holter. 1940. Beiträge zur enzymatischen histochemie. XXXIII. Die Atmung animaler und vegetativer Keimhälften von *Paracentrotus lividus*. Compt. Rend. Lab. Carlsberg, Sér. Chim., 23: 257–288.

Lindahl, P. E., and Å. Lennerstrand. 1942. Über Cozymase in der Amphibienentwicklung. Ark. Kemi, 15B: 1–13.

Løvtrup, Søren. 1953. 1. Energy sources of amphibian embryogenesis. 2. Utilization of reserve material. 3. Changes in peptidase content. 4. Temperature and amphibian embryogenesis. Compt. rend. Lab. Carlsberg, Sér. Chim., 28: 371–462.

Mezger-Freed, Liselotte. 1953. Phosphoprotein phosphatase activity in normal, haploid and hybrid amphibian development. J. Cell. and Comp. Physiol., 41: 493–517.

Moore, J. A. 1941. Developmental rate of hybrid frog. J. Exp. Zool., 86: 405–422.

——— 1946. Studies on the development of frog hybrids. J. Exp. Zool., 101: 173–219.

——— 1947. Studies in the development of frog hybrids. II. Competence of the gastrula ectoderm of Rana pipiens female × Rana sylvatica male hybrids. J. Exp. Zool., 105: 349–370.

Ornstein, Norma, and John R. Gregg. 1952. Respiratory metabolism of amphibian gastrula explants. Biol. Bull., 103: 407–420.

Panijel, J. 1950. L'organization du vitellus dans les oeufs d'amphibiens. Biochim. et Biophys. Acta, 5: 343–357.

——— 1951. Métabolisme des nucléoprotéines dans la gamétogenèse et la fécondation. Actualités Scientifiques et Industrielles. Hermann and Cie, Éditeurs. Paris. 288 pp.

Pomerat, C. M., and R. Haringa. 1939. The effect of some inhibitors of carbohydrate metabolism on the development of the frog egg. Anat. Rec., 75 (Suppl. I): 135.

Shumway, Waldo. 1940. Stages in the normal development of Rana pipiens. Anat. Rec., 78: 139–147.

Spiegelman, S., and F. Moog. 1945. A comparison of the effect of azide and cyanide on the development of frogs' eggs. Biol. Bull., 89: 122–130.

Spiegelman S., and H. B. Steinbach. 1945. Substrate enzyme orientation during embryonic development. Biol. Bull., 88: 254–268.

Sze, L. C. 1953a. Changes in the amount of desoxyribonucleic acid in the development of Rana pipiens. J. Exp. Zool., 122: 577–601.

——— 1953b. Respiration of the parts of the hybrid gastrula Rana pipiens × R. sylvatica. Science, 117: 479–480.

——— 1953c. Respiration of the parts of the Rana pipiens gastrula. Physiol. Zool., 26: 212–223.

ACKNOWLEDGMENTS

THE AUTHOR acknowledges with gratitude the generosity shown by the following in making the publication of the Bicentennial Editions and Studies possible: The Trustees of Columbia University, the Trustees of Columbia University Press, Mrs. W. Murray Crane, Mr. James Grossman, Mr. Herman Wouk, and friends of the late Robert Pitney who wish to remain anonymous.

Bei Fragen zur Produktsicherheit wenden Sie sich bitte an:
If you have any questions regarding product safety,
please contact:

Walter de Gruyter GmbH
Genthiner Straße 13
10785 Berlin
productsafety@degruyterbrill.com